Design of human nutrigenomics studies

Design of human nutrigenomics studies

edited by: S. Astley and L. Penn

Wageningen Academic
P u b l i s h e r s

S. Astley
Institute of Food Research, Norwich Research Park, Colney, Norwich, United Kingdom

L. Penn
Human Nutrition Research Centre, Institute for Ageing and Health, Newcastle University, William Leech Building, Framlington Place, Newcastle-upon-Tyne, United Kingdom

ISBN 978-90-8686-094-4

First published, 2009

© Wageningen Academic Publishers
The Netherlands, 2009

Table of contents

1. Introduction

G. Önning¹, M.M. Bergmann² and J.C. Mathers³

¹Biomedical Nutrition, Pure and Applied Biochemistry, Lund University, Lund, Sweden; gunilla.onning@tbiokem.lth.se
²German Institute of Human Nutrition Potsdam-Rehbrücke, Germany
³Human Nutrition Research Centre, Institute for Ageing and Health, Newcastle University, Newcastle-Upon-Tyne, United Kingdom

The health effects of nutrients and diets have traditionally been assessed using a few physiological end-points within a range of study designs. The omic technologies offer great promise for reaching a deeper understanding of the mechanisms through which food influences health (Müller and Kersten, 2003). Transcriptomic techniques can be used to measure RNA expression across the human genome in a single run and, although perhaps less well developed technologically, proteomics can be used to assess the relative abundance of thousands of proteins in cells and tissues. Metabolomics uses an array of technologies to measure the concentrations of a wide variety of low-molecular weight molecules in body fluids. In nutrition research, these tools can provide clues about the underlying mechanisms of beneficial and adverse effects of individual compounds as well as whole diets. In addition, omic technologies are being used to identify important genes, proteins and metabolites that are altered in the pre-disease state and thus are candidates for surrogate endpoints or molecular biomarkers of disease risk or susceptibility. An emerging aim of this nutrition research is personalised dietary advice, based on genotype in combination with a knowledge of phenotype and other lifestyle factors (e.g. physical activity), which may help reduce an individual's risk of developing common complex diseases such as diabetes and cardiovascular disease (Joost *et al.*, 2007).

The European Nutrigenomics Organisation (NuGO, www.nugo.org) is an EU-funded FP6 Network of Excellence (2004-2009). This publication is based on the NuGO workshop 'Design of human studies' held in Potsdam (Germany) during September 2008. The aim of this workshop was to review, and to provide examples of, successful human nutrition studies using nutrigenomics technologies. The workshop included presentations illustrating each of the major nutrigenomics techniques, namely transcriptomics, proteomics, metabolomics, nutrigenetics, and epigenomics.

Nutrigenomics is still a new science and the majority of studies have been carried out *in vitro* or in animal models. A few human studies have been published but recommendations on best practice in the use of nutrigenomics approaches in humans are still lacking. Although nutrigenomics offers several promising approaches, there are a number of obstacles and pitfalls in conducting, and interpreting, such studies in humans that add a layer of complexity to the challenges inherent in undertaking traditional nutritional intervention studies in healthy volunteers or patients. Amongst the more important is the question of sensitivity – i.e. the ability to detect effects of dietary intervention since these effects are comparatively small compared with those observed in, for example, pharmaceutical research. In addition, nutrigenomics studies in humans must deal with potential confounding factors from numerous sources including sex, body composition, age, health status, smoking, pharmaceutical and narcotic use, physiological and psychological stress, and other lifestyle choices (e.g. physical exercise), any of which may be difficult to quantify or control, and may add to 'noise' to the measurements. Thus, it is important, where possible, to collect data on all these factors and to develop protocols that minimise the impact of an individual changing their behaviour during the study, and confounding study outcomes. In some studies, it may be possible to standardise dietary intake but in longer-term studies such an approach is impractical and other strategies need to be adopted to reduce the impact of unwanted variation. Sampling needs to be standardised more rigorously than in other types of studies; diurnal variation can lead to a range of body-wide changes making it important to collect, for example, fasting blood samples at the same time on each of the study days. Overall, there must be careful control of the procedures governing how samples are collected, processed and stored.

One way to reduce the impact of inter-individual variability is to use a cross-over design in which each individual is their own control or recruit volunteers with the same (and known) genotype and/or phenotype. But, because of limited experience with human nutrigenomics studies, there is a limited evidence-base on which to base protocols more generally. For example, when should samples be taken after a meal to measure changes in the transcriptome, proteome or the metabolome. This is likely to depend on the aims of the study and the nutrient or non-nutrient (or food) under investigation since the kinetics of absorption, metabolism, and excretion differ in individuals and for different bioactive food compounds. Similarly, in longer-term studies, for how long should the intervention be applied to obtain stable changes in the transcriptome, proteome or metabolome. Should samples be taken at several time points to follow the time course, and if so how many samples and at what time intervals. In practice, the design of human nutrigenomic studies are a compromise between what the researcher would wish and what is financially possible (since most nutrigenomic techniques are expensive) in a limited number of individuals where the type and frequency of samples collected are also controlled by ethical considerations. The number of samples taken is constrained by the volume and complexity of nutrigenomics data produced because data analysis is both time consuming and difficult.

One of the major challenges in designing human nutrigenomics studies is the collection of appropriate body fluids or tissue samples. For practical and ethical reasons, such studies are often restricted to accessible body fluids such as blood (Van Erk *et al.*, 2006), urine or saliva (Walsh *et al.*, 2006). The transcriptome or proteome of peripheral blood mononuclear cells (PBMC) may change in response to a dietary intervention but, because it is not known whether there are comparable or different changes in other cells or tissues, such studies raise questions about the interpretation of data where the target tissue is, for example, liver, brain or bone. Where access to the target tissue is possible through biopsy, nutrigenomics studies can only be carried out using very small samples, as illustrated by the application of proteomics in human colo-rectal mucosal biopsies (Polley *et al.*, 2006), and invasive studies of this nature also raise ethical problems.

In this publication, use of the major nutrigenomic techniques in human studies is reviewed, and recommendations made, for best practise. These recommendations focus on design rather than issues around the use of the technology as this is discussed elsewhere.

Lydia Afman and her colleagues have experience in running both short-term nutritional studies as well as longer intervention studies with a focus on measuring changes in PBMC gene expression and adipose tissue after fasting and energy (i.e. caloric) restriction. Ivana Bobeldijk summarises the different factors influencing plasma and urinary metabolomes, and offers recommendations on the best study design. As in all omic studies, there may be added-value in using data from two or more separate studies, and Ivana also discusses the data management and analysis associated with pooling data from different studies. Lars Dragsted has used metabolomics to study the effects of apples consumption, and he presents the design, sampling procedure, and analysis used by his group. Lars also describes their experience in finding metabolomic markers for apple intake because there is considerable interest in the use of metabolomics as an objective measure of dietary exposure.

Since proteins are the major effectors in cells, proteomics has the potential to provide insights into how diet affects biological processes and identify novel biomarkers for health. Abigael Polley and Baukje de Roos review the different approaches for processing blood prior to proteomic analysis, including the removal of the most abundant plasma proteins, to allow investigation of the proteins of interest. They also consider approaches for isolating and measuring the proteome of specific circulating cell populations such as platelets. An outline of results from a study in which the analytical variability in 2-D gel electrophoresis was compared between four laboratories is presented by Baukje de Roos whilst Abigael Polley describes the use of proteomic analysis, by 2-D gel electrophoresis, in both blood and tissue samples. One of the major issues for proteomics is identification of specific protein spots because this may be influenced by factors such as genetic variability and post-translational modifications, which can change the isoelectric point a protein.

Anne Minihane has considerable experience in running human nutritional studies, and discusses important design considerations for these studies include genotype. Critical points include power calculations, ethical implications, matching genotype groups, and how to avoid introducing bias. Dietary exposure can also lead to chemical modifications of the genome, which are heritable across cell generations but do not involve changes in the primary DNA sequence. Such epigenetic marks include DNA methylation and modifications of chromatin. John Mathers introduces techniques that are used to investigate epigenetic markers, and discusses challenges involved in using these approaches in human nutritional studies.

In addition to issues about study design, the use of nutrigenomics approaches in human studies has ethical implications, which are more complex than those posed by more conventional nutrition research. An overview of the general ethical principles and a framework for human nutrigenomics studies is provided by the NuGO guidelines on bioethics (http://nugo.dife.de/bot/index.php), which was reviewed by Bergmann *et al.* (2008). The NuGO Bioethical Guidelines address four areas, namely (1) information and consent prior to a nutrigenomics study, (2) generation and use of genotype information, (3) establishment and maintenance of biobanks and (4) exchange of samples and data. As with all areas of research involving humans, these guidelines rest on the four principles of research ethics, viz. autonomy – i.e. self-determination, beneficence, non-malfeasance and justice where the balance of benefit and harm considers not only the individual but all groups in society. For example, in nutrigenetic research, researchers and ethics committees need to consider potential harm due to the violation of confidentiality or stigmatisation of an individual or group as well as the potential benefits from research involving genotypic information. One practical solution is to consent volunteers on the basis that genotype data will not be revealed to the participant or to any third party. But, the pooling of study data to test hypotheses with greater statistical power introduces additional ethical concerns, which are centred on the nature of the informed consent used originally in the collection of samples and associated data. Very narrowly worded consent may restrict or prevent secondary use, especially where samples and data are not unlinked-

G. Önning, M.M. Bergmann and J.C. Mathers

anonymised. Similar problems may also arise when setting up or using materials from biobanks for human nutrigenomics studies.

From the chapters in this book, it is clear that much progress has been made in the application of nutrigenomic technologies in human nutrition studies, and examples of best practice are emerging. However, there is a need for much more systematic investigation of design issues in human nutrigenomics studies if these technologies are to be applied effectively, and enable a deeper understanding of how diet influences health, and thus have an impact on the major diet-related public health issues of the 21st century, most notably obesity and healthy ageing. To help with this, NuGO has initiated a 'proof-of-principle study'. Human samples including plasma, platelets, PBMC, urine, and saliva have been collected at baseline and following a 36 hour fast, and samples are currently being analysed with transcriptomics, proteomics and metabolomics technologies in different laboratories across Europe. Collaboration is important in this area, and we need to continue to share data and experience to enhance the evidence-base for successful human nutrigenomic studies in the future.

14

Design of human nutrigenomics studies

2. Transcriptomics applications in human nutrigenomics studies

L. Afman and M. Müller

Division of Human Nutrition, Wageningen University, Wageningen, the Netherlands; lydia.afman@wur.nl

Transcriptomics

By employing high-density oligonucleotide microarray analysis, transcriptomics measures the set of messenger RNA (mRNA) in specific tissues or cells at the genome level. It is a comprehensive, sensitive and well-validated technique, which enables the expression of hundreds to thousands of genes, to be determined simultaneously, within a sample or across a study. As food compounds and nutritional status may have a profound influence on transcriptional regulation of genes, this technique can be applied to measure diet-induced changes in any tissue of interest (Müller and Kersten, 2003). A challenge in nutritional research, however, is the detection of relatively subtle effect of dietary intervention on cellular function and homeostatic control. The effects of diet on gene expression patterns are small and difficult isolate, making the development of highly sensitive, validated microarray analysis platforms for nutrition, and accurate methods of processing, essential.

Tissue sampling limitations

Although transcriptomics can be applied to any tissue, tissue collection is a major bottleneck in human studies. Transcriptional profiling requires accurate sampling of material for the extraction of sufficient, high quality RNA. In studies with patients who require surgery, this is less of a concern as biopsies from both diseased and surrounding healthy tissue can be obtained relatively easily. However, in studies with apparently healthy volunteers, which are common in nutrition research, access to tissues and organs is limited particularly internal organs such as the liver or pancreas, and visceral adipose tissue. One way to obtain such material from healthy individuals is to biopsy relatively accessible tissues including muscle and subcutaneous adipose tissue. Although other metabolic relevant organs (e.g. pancreas or liver) are not routinely accessible using this approach, transcriptional profiling of muscle and adipose biopsies is extremely useful as these organs fulfil important physiological functions, have a central role in energy metabolism, and are crucial in certain diet-related diseases including metabolic syndrome and Type 2 diabetes. Disadvantages of muscle and adipose biopsies are, however, the invasive nature procedure, and low yield of tissue and RNA. A less invasive alternative is blood cells as these can be obtained in relatively large numbers from sufficient numbers of individuals in a study. Interestingly, disease specific gene expression patterns have been identified using blood cells (Martin *et al.*, 2001; Valk *et al.*, 2004) and, although inter-individual variety in gene expression is high, low intra-individual variation means blood cells are also suitable for dietary intervention studies (Whitney *et al.*, 2003; Radich *et al.*, 2004; Cobb *et al.*, 2005).

Research question

The most important factor in a successful nutrigenomics study is the rationale supporting the study. The enormous amount of data generated in this type of study, in the absence of a clear research question, leads to a so-called 'fishing' exercise; a simple and robust hypothesis should be posed to avoid this. Moreover, it should be clear that the research question can only be answered by using whole genome transcriptomics rather than other methods, which are cheaper and less complex.

Appropriate study designs

In contrast to animal studies, which are generally performed in genetically-identical inbred mice strains, human studies involve a free-living heterogeneous population. Diversity in the genetic background of humans creates a large amount of variation in transcriptional profiles, which needs to be considered at the planning stage of a study. Several studies have examined variation in the transcriptional profiles of peripheral blood mononuclear cells (PBMC), and shown that inter-individual variation in these cells is much larger than intra-individual differences (Whitney *et al.*, 2003; Radich *et al.*, 2004; Cobb *et al.*, 2005; Eady *et al.*, 2005). This knowledge is important for the design of nutrigenomics studies, which are likely to elicit only small changes in gene expression. The large differences between individuals can be obviated by analysing responses within the same person, before and after dietary intervention. However, large cohorts are necessary if the intervention groups are independent, unless a cross-over design is used and each individual receives all treatments, which facilitates comparison of treatments within a person and increases the power of the study.

Within nutritional research a variety of intervention designs can be applied, which vary in length from a few hours in postprandial challenge studies to weeks in dietary studies or years in long-term interventions trials. The type and duration of the intervention is expected to have a substantial impact on changes in transcriptional profiles. However, whilst postprandial challenge studies may demonstrate the direct effect of a nutrient(s) on transcription profiles, long-term interventions are anticipated to induce changes in transcription, which will also reflect systemic alterations. The latter can be explained by the effect long-term dietary intervention will have on gene expression and metabolically active organs, such as the intestine and liver. These types of studies are of pivotal importance in increasing knowledge and understanding of the ways in which nutrients execute their effects in the body, both in the short- and long-term.

Planning and execution of intervention studies

Several other factors need to be taken into account when conducting human nutrigenomics research in addition to the overall design of the study. In order to be able to draw out causal relationships, between diet and changes in transcription, dietary variation should be limited within a study. Investigation of isolated nutrients allows the interpretation of data that would otherwise be far more difficult to understand. However, food is not ingested as single nutrients, and the combination of several bioactive compounds within the food matrix will influence digestion and uptake as well as transcriptional profiles. It can also be expected that the nature of any change(s) in transcription profiles is dependent on the types of dietary intervention. Thus far, studies that have shown changes in gene expression have been performed under acute nutritional condition such as fasting or caloric restriction (Bouwens *et al.*, 2007; Crujeiras *et al.*, 2008). But, knowledge about how long-term dietary interventions might be best performed in order for systemic changes to be measured properly is deficient. Moreover, intervention studies include metabolic challenges, such as glucose-tolerance or lipid-loading tests, which also influence transcription. Information about the time required for the body and transcription profiles to return to baseline is lacking, but highly relevant in the planning and execution of nutrigenomic studies.

Other aspects are known to influence gene expression; for example, exercise is known to modify gene expression in muscle but can also alter the PBMC transcriptome. Thus, exercise should be kept constant throughout a study as well as immediately before any samples are collected. In practice, this means participants must travel to a research facility the same way on each occasion and levels of exercise should be broadly similar throughout the study.

Gene expression is also altered by our circadian rhythm suggesting that gene expression profiles may vary throughout the day (Ptitsyn *et al.*, 2006; Zvonic *et al.*, 2006). This illustrates the importance of collecting samples from individuals at the same time of day in order to avoid changes that are induced solely by circadian rhythm. This is particularly important during postprandial studies, which are performed over the course of a day, where changes in gene expression may be caused not only by the intervention but also circadian rhythm. Factors that influence immune response such as infection, particularly the common cold and influenza, should be recorded in studies using immune system cells including PBMC. Similarly, antibiotics and vaccinations in the weeks before or during a study may affect an individual's response or the composition of a subpopulation of their PBMC and thereby the outcome of the study. As diet has a much smaller impact on gene expression than pharmaceuticals, prescribed and over-the-counter medications should be excluded during a study, although this is rarely feasible when working with some patient groups. Adipose tissue and muscle biopsies in postprandial and challenge studies should ideally be taken from the same position, particularly in the case of muscle as most people have a dominant leg, since there is a risk of inflammation and/or tissue repair in the area surrounding the biopsy, which will affect gene expression locally. However, it should also be remembered that the time between sessions can impact study outcome as well.

Because nutrients can influence transcription, samples should be taken from fasted volunteers, but in postprandial studies, it is also advisable for the previous evening's meal to be standardised to avoid cross-over effects, within an individual, which are associated with different meals.

Whilst these factors are vital to ensure reliable data, without a sufficiently large [, the power of the study will be too low for valid results. Variability in genetic background, transcriptional response and the subtle changes induced by dietary change mean a large cohort is essential in transcriptomic studies (Bouwens *et al.*, 2008). As in physiological studies, the number of individuals depends on the type of intervention and study design, and lower numbers are inevitably with cross-over designs or more extreme dietary intervention. The heterogeneity of humans also discourages the use of sample pooling in human transcriptomic studies. However, greater numbers of biological replicates will enhance accuracy by lowering the false-positive rate producing more reliable data (Kendziorski *et al.*, 2005).

Once a transcriptomic study has been performed, RNA quantity and quality need to be verified. The quality of the RNA is largely dependant upon sampling and RNA isolation methods. Ideally, subsequent labelling and hybridisation of RNA should be conducted by the same individual to minimise variation between arrays. Furthermore, samples from one person within a cross-over design should be labelled and hybridised in the same run to reduce variation induced by the procedure. Pre-processing of data requires careful attention as small changes in gene expression derived from dietary studies require a highly reliable algorithm (Irizarry *et al.*, 2005). But, the most challenging part of the process occurs once the array data have been obtained as the results must be assessed in terms of their biological relevance. Several commercial and non-commercial tools have been developed to assist with the extraction of significant changes and the visualisation of pathways or related networks (Curtis *et al.*, 2005).

Future perspectives

With the right approach, nutritional transcriptomic studies produce valuable information about how nutrients affect the human body and work at the cellular level. The next step in this research is the use of gene expression profiles for the identification of biomarkers, which are characteristic for the pre-disease state and will enable dietary intervention to prevent development of disease (Afman and Muller, 2006). Identification of an 'out-of-balance' system requires accurate phenotyping, which can be addressed with the application of metabolic challenge(s) to study the resilience of the body to cope with metabolic change. More comprehensive phenotyping can be achieved by combining transcriptomics and challenges tests with metabolomics and proteomics. With such information, better knowledge-based dietary advice for healthy eating can be given to groups in a population including elderly people, pregnant women, athletes, children, etc.

3. A metabolomics study on human dietary intervention with apples

L.O. Dragsted[1], M. Kristensen[2,3], G. Ravn-Haren[1,3] and S. Bügel[1]

[1]*Department of Human Nutrition, University of Copenhagen, Denmark; ldra@life.ku.dk*
[2]*Department of Food Science, University of Copenhagen, Denmark*
[3]*Danish Food Institute, Danish Technical University, Denmark*

Metabolomics is a promising tool for searching out new biomarkers and the development of hypotheses in nutrition research. Many different approaches to metabolomics exist; the ultimate goal being to make a quantitative fingerprint of all compounds present in a sample within a single analytical batch. While this goal is still far ahead, many researchers perform untargeted analyses using NMR or LC-MS to search for patterns or new markers, which may consist of some unknown or previously unconsidered compounds, or subtlety altered patterns of metabolite concentrations. In order to search for exposure and/or effect markers human or animal studies may be used, but only studies with an appropriate design will allow the effect of specific diets or dietary components to be determined. This chapter will discuss the design of human dietary intervention studies where samples are collected for metabolomics analyses as well as the analytical issues and data interpretation.

Design of human metabolomics studies

Design considerations for metabolomic studies are many and include the overall structure of the study as well as the control group, sample collection and handling, sample storage, sample analysis, QC measures and plans for data analyses. It is important to start with the final issue – the outcome measures. If disease outcomes are involved, parallel prospective, retrospective or cross-sectional designs are usually the only options. Metabolomic profiling of cross-sectional studies is problematic due to the inherent variability within a group. Such a study would need to be performed in a large, homogenous population. However, there is considerable risk that any diet-related effects might be lost in the noise from all kinds of genetic and lifestyle-related influences on the metabolome. In retrospective or prospective studies, with patients, the control group has to be selected with utmost care. A matched group with a different diagnosis would not be ideal since they would have characteristics unrelated to the targeted disease outcome, which are likely to influence their metabolome. A prospective design and a matched random population group would instead be preferable.

In longer-term prospective dietary trials a parallel design is, generally, the only feasible approach. Several samples should be collected during the study period in order to allow time-related analyses, and to eliminate noise associated with individual differences (Bijlsma *et al.*, 2006). This also allows trajectories to be drawn out at the individual and group levels.

Primarily, a meal study may be seen as a challenge study. For oral glucose and lipids, this is well documented but for other foods this is only now being realised as we are able to investigate the dietary perturbations of the metabolome, proteome and transcriptome. Carefully designed meal studies, using well characterised dietary compounds and whole foods with time-series collection of samples, can provide information about general and specific reactions to food components and whole diets. The description of a basic or reference diet for such studies is also an important part of the study design. This basic diet should be consumed for one or two days prior to the meal study. Whenever possible, cross-over dietary intervention studies are preferable because they offer better control of individual variation. This is particularly troublesome in metabolomics where each person has a range of genetic and environmental factors that cause specific patterns of compounds to emerge (Draisma *et al.*, 2008). Controlling these factors is only possible when repeat sampling is performed for each volunteer, preferably under each experimental condition; a strategy that is not possible in parallel studies.

Meal studies may also be performed using a cross-over design; the basic diet can be continued as a control meal during the study period. In a recent (unpublished data) human study with an onion preparation, we used a combined meal/cross-over design giving a whole day's dose as a single meal on the first day with frequent sampling up to 24 hours, and the volunteers continued with doses, spread over several meals, each day for the remainder of the week. In another study with a fruit and vegetable intervention (Dragsted *et al.*, 2006), we collected weekly samples, over a 25-day period, in addition to those taken at the beginning and the end, to allow for intra-individual time-series analyses of selected biomarkers.

Control of the diet

Since changes in diet strongly affect the metabolome (Walsh *et al.*, 2006), it is desirable to have control over meals, beverages and snacks for each volunteer. Quantities can be calculated using special kitchen facilities and experienced staff, and enable good control of dietary interference from factors unrelated to the intervention. However, a lack of resources for such facilities means many studies are undertaken with only limited dietary control. For example, only the test meals are given to the volunteers and/or recipe books using foods permitted during the study. In general, the less control there is over diet, the more noise affects the metabolome. If the contrast between experimental and control diets is large or clearly defined (e.g. a characteristic food item) noise may be less of an issue. Control diets also have to be considered carefully; for instance, what is the right control for wheat or oranges. These questions have to be answered for each study, based on the hypotheses proposed.

Sampling

Again, with respect to sampling, the goal of the study must be considered first. If the aim is to find new dietary exposure markers, sampling strategies should consider primarily the time course following dietary exposure. If the purpose is to identify markers to be used subsequently in other studies (e.g. epidemiology), it is important to consider the types of sample likely to be available from such studies. Typically, they could be fasting plasma samples and spot urine samples, in which case, studies should be designed to identify robust markers that would be appropriate using such samples. Spot urine samples are known to vary considerably due to both diet and diurnal variation (Slupsky *et al.*, 2007), and 24 hour urines are preferable in most studies.

Handling is a major source of variation in metabolomics (Dumas *et al.*, 2006). Ideally, samples should be processed and chilled quickly, and stored at -80 °C until analysis. However, samples are often processed by autosamplers, which typically means samples waiting in line for up to 48 hours at 5 °C prior to analysis. Markers that are highly sensitive to temperature would not readily survive such conditions, but most studies have to date suggested that plasma and urine can be analysed under these conditions without significant changes (Saude and Sykes, 2007). Faecal samples may also be useful, particularly in studies of prebiotics, but most laboratories avoid these samples because of the handling difficulties. However, simple methods for the collection and handling of faecal samples are available, and the samples are readily profiled by metabolomics. Examples of standard operating procedures (SOPs) for sample collection and handling of plasma, urine and faeces can be found on The European Nutrigenomics Organisation (NuGO) website (www.nugo.org).

PABA (p-aminobenzoic acid)

Urine is highly variable throughout the day, making 24 hour urine collections preferable for most studies. To ensure the completeness of urine samples, PABA (p-aminobenzoic acid) is often administered to the volunteers in three 80 mg doses during the 24-hour period. PABA is excreted as three metabolites; p-acetylaminobenzoic acid (m/z 179.0582) and p-acetylaminohippuric acid (m/z 236.0797) being the two major ones. Both are readily observed by LC-MS/MS analysis in the negative mode, and can be detected in positive mode. They serve as compliance markers, but once quantitative excretion has been assured they are also useful as internal standards for profiling analyses using LC-MS. Since the former has a mass equal to hippuric acid, and a similar retention time, care has to be taken to use a system that can obtain information about both metabolites and allows them to be distinguished from one another. If postprandial plasma samples are collected during PABA dosing care also has to be taken during their subsequent analysis. Faecal samples represent a particular problem; food is mixed in the gut and transit times vary from less than 24 hours up to a week or longer. The most recent specimen should be used for metabolomics. The brownish bile pigment, stercobilin, is readily observed in both the positive and negative modes, and may serve as a marker of the relative amount of faeces analysed, but this has yet to be confirmed.

Analytical strategy

The analytical strategy to test the primary hypothesis is an important part of any study plan, but it may open up opportunities for additional post-hoc exploratory analyses using different metabolomic profiling and fingerprinting methods.

We use untargeted LC-TOF and NMR fingerprinting analyses as well as targeted methods. Regardless of method, however, it is important to include standard samples. A set of internal standards for addition to biological sample before profiling has yet to be described, but might become essential in the near future. A set of external standards and appropriate blinding as well as sample specific standards (i.e. standard urine or plasma samples) are used by most laboratories. External standards are analysed at regular intervals in each analytical batch, and allow between- and within-run quality control of parameters such as spectra, mass accuracy and retention time stability.

Randomisation of sample order is crucial to metabolomics since analytical drifts are unavoidable but randomisation also depends on the study design. In cross-sectional and parallel studies, samples from all groups should be analysed in random order. However, repeat samples from the same person should be analysed in the same batch to allow for control of intra-individual variation. This is even more important in cross-over intervention and time-series studies, where the hypotheses are concerned within-person variation. Within a batch, it is important to analyse the samples from each person at random to avoid systematical errors. Each sample should be analysed repeatedly, preferably in different batches, if multi-batch analysis is necessary. External standards, within each batch, are useful for controlling batch-to-batch analytical variation. We use a 40-compound external standard for metabolomic analyses of all sample types, and the SOP for producing this standard has been included on the NuGO website.

Identification of biomarkers

There are many software options to view and align metabolic profiles, but this issue is beyond the scope of this publication. Nevertheless, it is important to align profiles according to the performance of the analytical equipment. For example, if retention time drift is short, binning (assessing) across large intervals gathers several compounds into the same bin in a median retention time. Software must allow ready inspection of the chromatograms or spectra and the collected markers or features as a minimum requirement. Each chromatogram/spectrum must be inspected visually in order to remove those with intensities above or below the appropriate thresholds or other flaws discernible by eye.

Markers may be identified using a multitude of statistical strategies, which are also outside the scope of this paper. However, the markers should be visualised, graphically or mathematically, across the dataset to evaluate whether design flaws, analytical errors or other factors could have affected the results. Simply sorting according to analytical order, batch, person etc. gives information on analytical integrity immediately if the value of one or more markers in each sample is plotted against sample order. Samples or even batches should be rejected, and corrective measures implemented, if systematic errors are observed. Repeated analysis of each sample across different batches allows for rejection of some results without loss entire data sets.

A good marker shows a clear-cut change (i.e. increase or decrease) when data are sorted by group or time, and this relationship should not be affected by batch or order of analysis. Visual or mathematical inspection allows discrimination of unique (zero-to-something) markers, typical of metabolites derived from the dietary constituents under examination, from markers present in all samples. The latter, showing more limited responses are typical of endogenous metabolites and those derived from the gut flora, and consequently depict more functional changes. Functional markers are particularly interesting for more advanced analysis of co-variation with known variables or with markers of effect such as changes in gene expression, clinical chemistry or physiology. Further validation in additional studies is necessary to ascertain whether any marker has more general applicability.

Comparison studies

Meal studies are useful for validation of exposure markers. However, animal studies may also be useful since many of their metabolites are similar, and better control can be exercised; we have observed good concordance between exposure markers in studies with fruit intervention in animals and humans (unpublished data). Animal studies may also be useful in supporting validation of human exposure markers, and they have the added benefit of allowing more invasive samples to be obtained, which might be useful in elucidating more details about metabolic and functional effects. Animal diets can exclude specific dietary components to identify a dietary reference metabolome. This is more difficult in humans where semi-synthetic or synthetic diets are uncommon, although one or two days consuming astronaut food could be the starting point for a well designed reference human metabolomics study.

Example: a pilot analysis of samples from a cross-over intervention study with apples

In a recent cross-over intervention study with apples, 24 hour urine samples were collected at the end of a control period, following a diet free from fruit and vegetables, and again at the end of the intervention period with the same diet plus three apples per day. The samples were profiled on a UPLC-TOF system (QTOF Premier, Waters, Milford, MA, USA) over four minutes using a BEH C18 column, and data aligned using the Markerlynx software provided by the same manufacturer. The sample-marker matrix was normalised and Pareto-scaled before principal component analysis (PCA). Several of the principal components allocated the samples into two sets according to diet with almost complete separation. The six repeat analyses for each sample also clustered tightly. Trajectories were drawn for each volunteer between their control and apple period samples. The trajectories were horizontal for males and more vertical for the female volunteers, which allowed markers altered by apple consumption and their influence by sex to be identified from the loadings plot within the same analysis (Figure 3.1).

Conclusion

Metabolomics profiling and multivariate analysis can be used to search for exposure and effect markers in human dietary studies. Cross-over

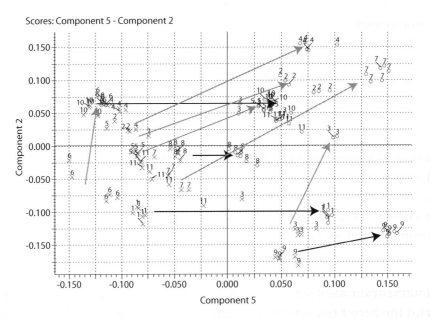

Figure 3.1A. Principal component analysis of a cross-over apple intervention study. Scores plot showing second and fifth principal component and clusters of hexuplicate analyses of each sample collected with (upper right) or without (lower left) apples in the diet. Arrows are trajectories for samples from each volunteer taken after each dietary period. Horizontal lines are from males whilst the more vertical lines are from female volunteers.

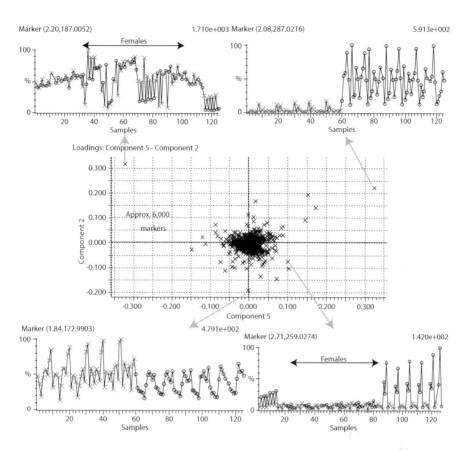

Figure 3.1B. Principal component analysis of a cross-over apple intervention study. Loadings plot and selected markers from the same analysis where the markers affected positively by apple consumption are to the upper right whilst those affected negatively are to the lower left. Markers affected differentially by sex are in the lower right and upper left, respectively.

intervention studies allow control of large inter-individual variations, and the better the control over diet, study protocol, sample handling and analytical procedures, the easier it is to single out changes related to diet.

4. Metabolomics in human nutrition studies

I. Bobeldijk-Pastorova[1], S. Wopereis[2], C. Rubingh[1], F. van der Kloet[2], H. Hendriks[2], E. Verheij[1] and B. van Ommen[2]

[1]*TNO Quality of Life, BU Quality and Safety, Zeist, the Netherlands; ivana.bobeldijk@tno.nl*
[2]*TNO Quality of Life, BU Biosciences, Zeist, the Netherlands*

The primary goal of nutrition research is to optimise health through dietary intervention to prevent or delay the onset of diseases and ageing. Clinical trial design is essential to ensure a meaningful outcome from a nutritional study. Long-term human intervention studies that measure the effect of a nutrient on biological endpoints, in relation to health promotion or disease prevention, are the golden standard in nutritional sciences. However, these studies are difficult to perform for experimental, economic and ethical reasons.

Application of advanced system biology tools, such as transcriptomics, proteomics or metabolomics, in human intervention studies increases the amount of information available, and may perhaps decrease the duration of the study. Metabolomics measures the relatively small molecules (endogenous and exogenous, <1 kDa) in tissues or biological fluids offering nutrition research an alternative to the single biomarker approach currently used to assess health and disease. As a technique, metabolomics can identify metabolites that may make the difference between the effects of different diets clear, and it can expand our knowledge of human health and the interacting and regulatory roles of nutrition. However, the use of such a powerful technique does not eliminate the necessity of proper study design. Appropriate (intervention) study and experimental (analysis and statistics) designs, which take in to account the limitations of the tools used, are essential to ensure data quality and derivation of useful information from these complex studies. In nutritional intervention studies, it is particularly important to be able to control and understand factors that contribute to variation in the data because they often deal with subtle changes in nutrition, which affect cell and molecular processes. Both normal physiological variation (i.e. biological variation) and analytical variation need to be controlled or minimised to avoid metabolic fluctuation being confused with changes in biomarkers, which represent a metabolic change associated with the nutritional intervention under examination.

This chapter describes study designs and approaches suitable for controlling variation due to sampling and metabolome analysis as well as the extent of biological variation observed in the metabolome. Some of these topics will be demonstrated using results from an example intervention study.

Example metabolomics study

Our nutritional intervention study involved 36 volunteers assigned to four groups. Each received four treatments, including a placebo, in a cross-over design (Bakker *et al.*; personal communication). Plasma samples collected during this study were analysed by GC-MS (Koek *et al.*, 2006), LC-MS free fatty acids and LC-MS lipids, respectively (Bijlsma *et al.*, 2006; Bergheanu *et al.*, 2008).

Study design

Currently, two designs are commonly applied to dietary intervention studies, which are also suitable for metabolomics analysis; a parallel design and a cross-over design. In the parallel design at least two groups of people are followed-up over time, and one or more groups receive a treatment whilst another acts as a control. The intervention effect is defined as the difference between selected parameters at the start and the end of the intervention period compared with the control group. In a cross-over design, each person undergoes all treatments sequentially including (ideally) a control period. The intervention effect is defined as the difference between selected parameters at the end of an intervention as compared with the same parameter at the end of the control period. Each study design has advantages and disadvantages, which have been described elsewhere (Coulier *et al.*, in press). For metabolomics, specifically, either is equally suitable and the choice depends on the hypothesis or research question under examination.

Both the cross-over and the parallel designs can be extended to include an acute homeostatic perturbation where, for example, volunteers are challenged with a glucose tolerance test. Although this concept is new in nutrigenomics research (B. van Ommen, personal communication) challenge tests are common in medical sciences (e.g. diagnosis of insulin resistance uses a glucose tolerance test). This experimental design uses one or more stressors in such a way that an individual's homeostatic control is safely and reversibly challenged. Application of a metabolic change, in combination with metabolomics, may be helpful in identifying a set of metabolites that are predictive of differences in response to the challenge, and in this way help describe a 'more healthy' state (B. van Ommen, personal communication; Shaham *et al.*, 2008; Wopereis *et al.*, 2009).

Cost is often a limiting factor in executing the perfect study. Most studies have a restricted number of subjects, and power analysis can be used to estimate the minimum number of samples needed to ensure the experiment can successfully discriminate between treatments. However, the literature on power calculations for study design, which include metabolomics, is very scarce. In most cases, univariate approaches are used and the multi or megavariate character of metabolomics data is not taken into consideration (Coulier *et al.*, in press).

Between 2007 and 2008, papers emerged that discussed conceptually the different study designs where n=1 for personalised medicine and dietary intervention (Schnackenberg *et al.*, 2008; Van der Greef *et al.* 2006). In this context, personalised medicine is defined as customised medical care for each patient's unique condition (Schnackenberg *et al.*, 2008), and metabolomics as the ideal technology platform for the discovery of biomarkers, which can be used in personalised health monitoring programmes and the design of individualised intervention (Schnackenberg *et al.*, 2008; Van der Greef *et al.*, 2007). Examples are scarce and opinions about the design and interpretation of resulting data are still divided; these 'personalised' studies will, however, not be considered here.

All study designs have one thing in common; at least two sampling points for each person at the beginning (t=0) and the end of study (t=x, per treatment in a parallel design or t=end for cross-over designs). Although (multivariate) data analysis is one of the later steps, it needs to be considered when designing a study as the structure of the design should be included in the statistical modelling. For example, in a study where healthy individuals are compared with patients, and the development of metabolic changes are to be traced across time, the structure of the design needs to be included in the data analysis to ensure the sources of variation are not confused. Including the experimental design in the model assessment helps to ensure the results can be interpreted more easily.

While principal component analysis (PCA) and partial least squares (PLS) are well suited for two-way datasets, structured data need to be processed with methods such as analysis of variance (ANOVA)-simultaneous component analysis (ASCA) and n-PLS (multiway-PLS), which can deal with more than two sources of variation in the data. Several examples of the application of structured datasets in nutritional studies have been described in the literature but are not discussed here (C.M. Rubingh, personal communication; Coulier *et al.*, in press; Smilde *et al.*, 2004; Thissen *et al.*, 2009).

Sample collection and handling during a study

The chemical diversity of the metabolome is enormous and spans a large, dynamic concentration range. A wide variety of methods have been used to separate and quantify the component metabolites (Bedair *et al.*, 2008; Issaq *et al.*, 2008; Gika *et al.*, 2008), but no single analytical platform can capture all the metabolites in a single sample. Thus, a technology platform comprised of several analytical approaches, based on different techniques, currently offers the best solution. The choice of analytical platforms, however, needs to be considered during development of the sample collection protocol.

The use of anti-coagulants can have adverse effects on some analytical platforms. GC-MS chromatograms from plasma, where heparin and citrate were used as anti-coagulants, are shown in Figure 4.1. Because citrate is derivatised by the method used, it is the most abundant peak in the chromatogram, and the detector has to be switched off during elution of this compound to avoid overloading it. As a consequence, information about all the metabolites eluting in the region as citrate is lost. Examples of base peak HPLC-MS chromatograms of plasma polar compounds, after butylation, are shown in Figure 4.2. Collected from the same individual, in blood tubes using heparin and EDTA as anti-coagulants, differences in abundant peaks can be observed. The EDTA-blood chromatogram contains a number of peaks formed by

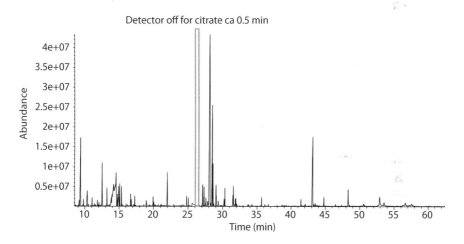

Figure 4.1. GC-MS TIC chromatogram of plasma collected with citrate. Samples were collected from 36 volunteers assigned to four groups, each receiving four treatments including a placebo in a cross-over design. Blood was collected using either heparin or citrate, and plasma samples compared using GC-MS. During elution of citrate, the detector is switched off to avoid overloading it. As a consequence, information about 12 metabolites eluting in the same region is lost. Examples of compounds that might be missing include: N-acetyl-DL-glutamic acid (2TMS and 3TMS), L-glutamine, DL-glyceraldehyde-3-phosphate (peaks 1 and 2), L-histidine (1TMS), sn-glycerol-3-phosphate, 5(4)-aminoimidazole-4(5)-carboxamide (3TMS and 4TMS), 4-hydroxymandelic acid, 5,6-dimethylbenzimidazol (1TMS), and hypoxanthine.

derivatisation of EDTA. These peaks not only distort the chemical analysis (i.e. matrix effect on co-eluting peaks), but also affect the statistical analysis when the tubes are not filled with the same volume of blood because of the differences in plasma-EDTA concentrations. In general, because citrate and EDTA interfere with analysis of polar compounds from plasma in all methods, heparin is the preferred anti-coagulant as it is not derivatised and, therefore, does not interfere with analysis. Furthermore, no differences have been observed between EDTA- and heparinised plasma lipid profiles, where no derivatisation is involved.

For some metabolites it may be important to stop enzymatic activity after sample collection. This can be achieved with the addition of enzyme inhibitors in plasma or conservation agents such as azides in urine, or appropriate collection (snap freezing) and storage (-70 °C). In large studies, when many samples are collected simultaneously for different analyses, it can be difficult to process them immediately. In such cases, it is important that samples destined for the same analysis are treated in the same way to avoid the introduction of artefacts. Handling-induced biological variation, in particular, can be limited by ensuring any storage of all samples is delayed for 15 minutes rather than each being stored individually.

Controlling analytical variation

In addition to appropriate sample handling, analytical variation caused by poor performance of the selected platform and instrumental drift are major problems in metabolomics. Method performance for individual metabolites depends on their chemical properties and concentration. Instrumental drift is particularly important in nutritional studies where a large number of samples have to be analysed over a series of batches. If instruments are cleaned between measurements, systematic differences in the data between batches are more likely, although the severity depends on the platform and the chemical properties of the metabolite. Internal standards and randomisation as well as proper control of the experimental settings can do much to minimise analytical variation. But, in some cases, the combination of analytical and biological variation

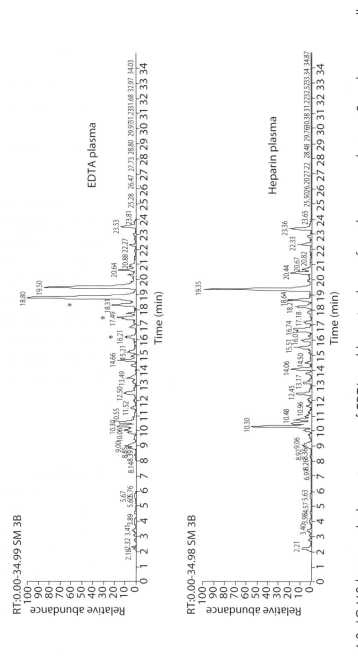

Figure 4.2. LC-MS base peak chromatograms of EDTA and heparin plasma from the same volunteer. Samples were collected from 36 volunteers assigned to four groups, each receiving four treatments including a placebo in a cross-over design. Blood was collected using either heparin or EDTA, and plasma samples analysed by LC-MS. In plasma from the same individual, polar compounds (i.e. amino acids, small organic acids, amides, small peptides) were derivatised with butanol prior to LC-MS analysis. Peaks indicated with * are EDTA derivatisation products.

may be so large that they prevent detection of changes induced by the nutritional intervention.

The use of pooled study samples for quality control (QC) has recently been described in the literature (Van der Greef *et al.*, 2007; Sangster *et al.*, 2006). A pooled study sample is analysed within each analytical batch. It reflects the average metabolite concentration across the study and contains the same features as the other samples allowing the performance of the analytical platform to be assessed by calculating the coefficient of variation (CV) in the pooled samples. The same sample data can also be used for calibration and to remove any offset between or drift within the batches (Van der Greef *et al.*, 2007; F. van der Kloet, personal communication). Data correction (removal of analytical variation) is crucial for studies of this magnitude as demonstrated by our example study; more than 500 samples from the 36 volunteers were analysed by GC-MS generating almost 900 hours of analysis following 663 sample injections. Each of the 19 sample batches included QC samples, and QC correction significantly improved the quality of the data, which is shown in Table 4.1. Correction for the response of internal standards removed much of the systematic offset amongst the batches. However, for some metabolites, the internal standard is not sufficient to remove all of the offset and drift. After QC correction, the quality of the data is further improved and, in some cases, this can be enough to detect the subtle differences amongst the groups.

Figure 4.3 shows two metabolites from our example study where the total variation amongst the treatment groups are shown as box-and-whisker plots after normalisation for the internal standard, and after normalisation for internal standard and QC correction. The analytical variation was reduced significantly after QC correction, following internal standard normalisation, and the remaining variation is to a large extent biological variation.

Despite all the quality control measures, instrumental failure or human error still can occur. If too little sample is available for re-analysis part of the dataset will be lost, which means randomisation is also important especially in large-scale nutritional studies. This does not mean that

Table 4.1. Example of improved data quality using internal standards and quality control samples in a GC-MS dataset. Samples were collected from 36 volunteers assigned to four groups, each receiving four treatments including a placebo in a cross-over design. More than 500 plasma samples were analysed by GC-MS in 19 batches, which also included quality control (QC) samples. Correction for the response of internal standard removes much of the systematic offset amongst the batches (B) but QC correction (C) improved the quality of the data further.

Analytical CV	Raw data	Normalised against internal standard	Normalised against internal standard and pooled QC samples
<10%	37	84	123
10-20%	46	35	21
21-30%	39	14	1
>30%	23	12	0
Total analytes	145	145	145

samples are necessarily randomly selected for analysis, but rather that careful consideration is given to the order of analysis, based on the study design. For example, in a study involving a homeostatic challenge, an important outcome of the metabolite profiling is the area-under-the-curve for individual metabolites in response to the challenge. In such studies, samples from the same individual should be in a single batch, randomised on the basis of time. In other studies, it may be more important to randomise all of the samples across each of the analytical batches.

Controlling biological variation

In most nutritional studies, plasma or urine is the matrix of choice as either can be collected relatively easily compared with tissue (liver, adipose or muscle). In addition to the technical parameters that can influence the results, however, sample preparation, choice of instrumentation and instrument settings as well as differences amongst the samples

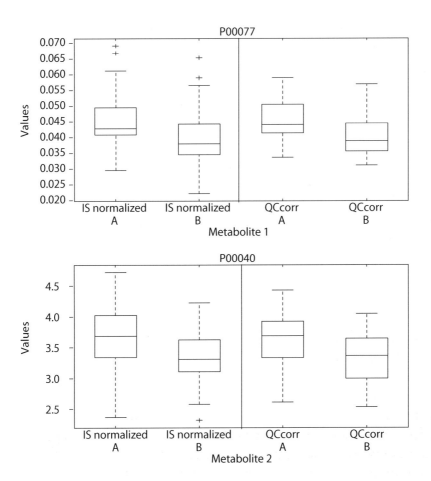

Figure 4.3. Examples of box-and-whisker plots of two selected metabolites analysed by GC-MS in fasting plasma after treatments A and B. Samples were collected from 36 volunteers assigned to four groups, each receiving four treatments including a placebo in a cross-over design. Plasma samples were analysed using GC-MS, and data presented following normalisation against the internal standard (left-hand side for each metabolite) or the internal standard and quality control pooled samples (QC correction) (right-hand side for each metabolite). The (analytical) variation within the groups is reduced after QC correction (i.e. whiskers are shorter).

(biological variation) can influence the accuracy of data collected. Large differences in sample composition can affect the response of one or more metabolites by, for example, ion suppression effects in electron spray ionisation (ESI) or impaired chromatographic performance in GC-MS, and lead to an erroneous response for co-eluting metabolites (Van der Greef *et al.*, 2007). Variations in the concentrations of highly abundant substances (e.g. glucose and urea) also have an effect on responses for other metabolites. Typically, this is not a problem in fasting plasma samples because inter-individual sample composition is within the same order of magnitude, despite the biological variation. However, the concentrations of many other less abundant metabolites can vary significantly amongst individuals, regardless of treatment, or within an individual because of dietary intervention.

Examples of biological variation

As a measure of biological variation, coefficients of variation were calculated (i.e. the ratio of the standard deviations of the mean for each metabolite detected against their concentration in all fasting plasma samples). Depending on the platform selected, two or four fasting plasma samples were analysed for each person. Table 4.2 shows the number of metabolites in each quartile. The ratios between the maxima and minima concentrations were also calculated for each metabolite as an additional measure of the biological variation. The upper portion of the table (A) shows the variation observed within this study including treatment whilst the lower portion (B) includes only the biological variation observed in the control group.

Using GC-MS, 145 separate metabolites were detected. The biological variation in relative metabolite concentration for most (143 out of 145) was less than 50%. Nineteen out of 145 metabolites displayed at least an order of magnitude difference between the highest and lowest relative concentrations. For free fatty acids, as determined by LC-MS, only four (of the 22 detected) fatty acids showed more than an order of magnitude difference between the highest and lowest relative concentrations. High biological variation was observed in some lipids measured using the LC-MS lipid method. On average, more variation was observed in

I. Bobeldijk-Pastorova, S. Wopereis, C. Rubingh, F. Van der Kloet, H. Hendriks, E. Verheij and B. Van Ommen

Table 4.2. Summary of the biological variation as observed for different metabolomics platforms within fasted plasma samples from our example metabolomics study. Samples were collected from 36 volunteers assigned to four groups, each receiving four treatments including a placebo in a cross-over design. Depending on the platform selected, two or four fasting plasma samples were analysed for each person, and the number of metabolites in each quartile presented. The ratios between the maxima and minima concentrations were calculated for each metabolite as an additional measure of the biological variation. The upper portion of the table (A) shows the variation observed within this study including treatment whilst the lower portion (B) includes only the biological variation observed in the control group.

			GC-MS	LC-MS FFA	LC-MS lipids
A[1]	Total number metabolites		145	22	104
	Biological CV[1]	<20%	21	0	18
		20-50%	106	20	40
		51-100%	16	2	35
		>100%	2	0	11
	max/min[3]	<5	67	8	26
		5 to 10	59	10	28
		>10	19	4	50
	Fasting plasma per volunteer[4]		2	4	4
B[2]	Total number metabolites		145	22	110
	Biological CV[2]	<20%	55	4	22
		20-50%	78	18	44
		51-100%	8	0	28
		>100%	4	0	16
	max/min[3]	<5	122	18	48
		5 to 10	11	4	23
		>10	12	0	39
	Fasting plasma per volunteer[5]		1	1	1

[1] CV calculated over all biological samples (irrespective of treatment).
[2] CV calculated over samples from only the placebo group.
[3] Ratio between maxima and minima observed (relative) concentrations for each analyte.
[4] Fasting plasma taken from each individual after a different nutritional intervention.
[5] Fasting plasma taken from each individual only after the placebo treatment.

Design of human nutrigenomics studies

the lipid compounds measured (104 in total including triglycerides, lyso-phosphatidyl cholines, phosphatidyl cholines, cholesterol esters, diglycerides and sphingomyelins); about 40% had a biological CV greater than 50% and about the same portion were determined to have more than an order of magnitude difference between their highest and lowest relative concentrations. Although more metabolites fall into the lowest quartile (Table 4.2B) from the placebo group, more biological variation was observed in the treatment groups (Table 4.2A).

Biological variation within treatment groups can be a serious problem as changes in endogenous metabolites, induced by the nutritional intervention, are often smaller than inter-individual differences after fasting. Some of the biological variation observed in NMR spectra of urine can be removed by placing volunteers on a standardised diet for at least 24 hours prior to sampling (Walsh *et al.*, 2006). For plasma, however, this is not the case.

In most studies, subjects are carefully selected on the basis of strict inclusion and exclusion criteria, and in this way biological variation is controlled. Further control can be achieved by collecting samples from the same individual on more than one occasion during the course of the study as well as at the start (t=0) and end of the study. After chemical analysis and data processing, data for each analyte can be mean-centred for each person (Bijlsma *et al.*, 2006). This helps to reduce the inter-individual variation within a treatment group significantly, and enhances the otherwise subtle differences amongst the treatment groups. Figure 4.4 shows an example of plasma lipid data from our example study. Treatments A, C and D do not cluster separately if PCA is performed on auto-scaled data, and treatment B only shows partial separation due to the large biological variation in the plasma lipid profiles within the treatment groups. After mean-centring of the data for each individual, the difference between treatment B and the other treatments becomes clearer with almost no overlap.

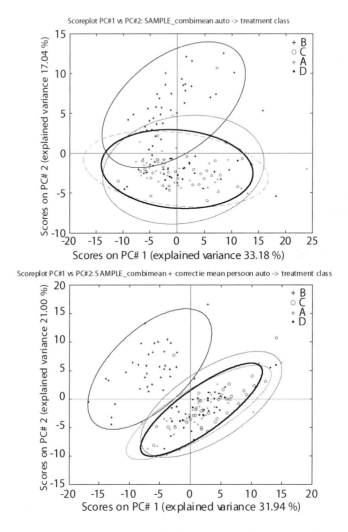

Figure 4.4. Lipids in fasting plasma of subjects. Samples were collected from 36 volunteers assigned to four groups, each receiving four treatments including a placebo (A-D) in a cross-over design. Plasma lipids were analysed using LC-MS; principal component analysis (PCA) was applied to auto-scaled data (left-hand side) or data that was mean-centred for each individual (right-hand side) prior to being auto-scaled. Treatments A, C and D do not cluster separately if PCA is performed on auto-scaled data, and treatment B only shows partial separation due to the large biological variation. After mean-centring of the data, the differences between treatment B and the other treatments become clearer with almost no overlap (elipse: 95 % c.i. mva).

Conclusions

It can be concluded that for a successful study including complex approaches such as metabolomics, collaboration between statisticians, dieticians, analytical chemists and biologists is essential during the design phase. In general, well established study designs such as cross-over and parallel studies are suitable for nutritional metabolomics studies. However, in addition to the intervention design and statistical analysis of data, appropriate analytical approaches, randomisation of sample analysis, and adequate quality control are also important. Only then is it possible to reduce the analytical variation and control the number of artefacts sufficiently to enable researchers to contend with the challenge of biological variation in response to dietary intervention.

5. Proteomics and the trouble with people

A.C.J. Polley, I.T. Johnson and F. Mulholland

Institute of Food Research, Norwich Research Park, Colney, Norwich, United Kingdom; abigael.polley@bbsrc.ac.uk

Proteomics has largely been used as a comparative technique, in the sense that changes in the ratio of specific proteins are compared under different conditions. Unlike most other omic studies, proteomics requires the identification of the proteins as part of the investigation, and most of the focus within studies has been on discovering which proteins have changed. Unlike conventional microarrays, however, where one spot is equal to one gene, proteins can exist in several isoforms derived from a single gene, and these different products have biological significance. Simply identifying the proteins that have changed, therefore, does not always tell the whole story.

Two-dimensional (2-D) gel electrophoresis is used widely in proteomic studies. It is an extremely powerful technique, which separates proteins using their intrinsic charge and size, to reveal differences within the individual proteins, which indicate the presence of post translational modifications (PTMs). It is, however, limited by the types of protein that can be analysed; important membrane proteins are problematic to extract and low abundance proteins are difficult to analyse due to restrictions in the dynamic range. The technique is labour-intensive, and can be subject to wide technical variance, which means generating statistically significant data requires great emphasis on good laboratory skills and practice.

In human studies, one of the major problems with proteomics is genetic variability, whereby one or more spots (proteins) appear or shift location on a 2-D gel. There are several potential reasons for this. For examples, single nucleotide polymorphisms (SNPs) can cause a shift in the isoelectric point of a protein, thus affecting its location on a 2-D gel. Secondary effects such as protein modifications (e.g. glycosylation of certain amino acids within the protein) can also cause a shift as well as the formation of multiple species of the same protein. These factors need to be considered when designing a human nutritional study involving proteomics.

In any study using a comparative technique (as in 2-D gel electrophoresis) it is essential to choose the correct controls. Generally, in human studies, deciding upon and obtaining an appropriate control can be very challenging. For human nutritional studies, the only appropriate control may be the individual, with samples obtained such that the proteome can be assessed pre- and post-intervention. However, the amount of sampling required for statistically significant results may, in such cases, become an ethical issue. Ethical approval plays a large part in the design and execution of human studies, and ethics committees will need to be convinced of the merit of a study before allowing volunteers to be subjected to repeated invasive sample collection (e.g. tissue biopsies). Collection of biopsies often also requires commitment from surgeons, who are aware of the importance of appropriate sample collection techniques, in order to avoid the introduction of sampling artefacts to the data.

The use of less- or non-invasive techniques is preferable for human studies involving healthy volunteers or patients, but then sampling is restricted to whole blood components such as plasma, platelets and peripheral blood mononuclear cells (PBMCs). Although such samples may be more readily obtained than solid tissue biopsies, many studies still require samples from more direct targets. In studies comparing normal tissue with tumours, the control is usually tissue from the non-tumour region in the same cancer patient (Stulik et al., 1999; Friedman et al., 2004). In our studies (Polley et al., 2006), we have shown this apparently normal tissue, from patients with colon cancer, does not give the same proteomic profile as normal tissue obtained from patients with no colonic pathology.

Establishing a reproducible method for sample collection is a key element of any study. A standardised and robust method for collection of plasma, platelets and PBMCs from human blood is described in De Roos *et al.* (2008a), but these techniques can be applied to most samples. De Roos *et al.* (2008a) also describes the variability that can exist amongst laboratories applying 2-D electrophoresis to the same samples, and it is important that researchers are aware of the technical variability of their own analysis systems. A critical pathway analysis (Figure 5.1) of proteomics procedures (from sample collection to spot identification) demonstrates the number of points where variation can be introduced, and highlights the necessity of maintaining accuracy and precision to minimise the extent of technical variability in the system. Although low technical variability is important, Horgan (2007) compared technical variability with biological variation between individuals, using 2-D gel electrophoresis, and describes how adequate biological replication is also essential. The number of samples and replicates required for a human study is study-dependent, and Horgan provides power calculations for determining these numbers.

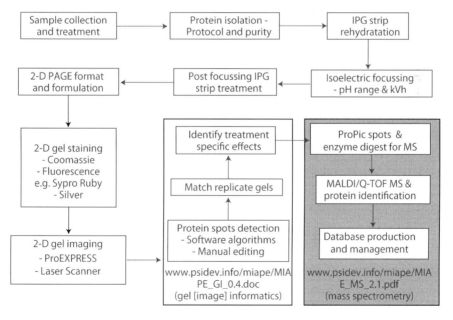

Figure 5.1. Critical pathway analysis of the proteomics procedure from sample collection to spot identification.

2-D gel reference map websites (e.g. www.expasy.org/ch2dothergifs/publi/plasma-acidic.gif and www.expasy.org/ch2dothergifs/publi/plasma-basic.gif) can be used to indicate the likely identity of a protein, but their value is limited because different inter-laboratory factors (e.g. use of isocratic versus gradient gels) can affect spot profiles (i.e. spots shift by molecular weight and pI). It is, therefore, essential that the identity of a protein is confirmed independently and, indeed, for publication purposes most journals require confirmation of the observed changes using other techniques (e.g. enzymatic activity or Western blotting).

Where possible, proteomics samples should be subjected to as little processing as possible as this can introduce variation due to either losses or modifications derived from the processing (e.g. proteolysis within the sample). The 2-D gel electrophoresis analysis of plasma, however, provides an example where some processing is usually essential. In plasma, two proteins, albumin and IgG comprise about 66% of the protein mass (50% and 16%, respectively). Another 10 proteins (in descending concentration transferrin, fibrinogen, IgA, α2-macroglobin, IgM, α1-antitrypsin, haptoglobin, α1-acid glycoprotein, apolipoprotein A-1, apolipoprotein A-2) constitute a further 30% with the less abundant proteins in the final 4%. To detect these proteins using 2-D gel technology, some of the more abundant proteins have to be depleted first. A number of commercial columns are available to remove highly abundant proteins (Table 5.1). Some of these remove up to 28 proteins, but care needs to be taken as these proteins may be biologically important and depletion inappropriate for the hypothesis under investigation. In recent work, we have chosen to use the single use GE Healthcare Albumin and IgG Removal kit (GE Healthcare Life Sciences, Amersham, UK), largely on the grounds of higher throughput and concerns over reproducibility associated with re-use of depletion columns for multiple samples. Cost is frequently important; not only the price of the kit but also the labour costs associated with carrying out the depletions.

Protein samples often need to be desalted and concentrated before the first dimension IEF (isoelectrical focusing) step. Historically, protein

Table 5.1. Some of the commercial columns available for removing highly-abundant proteins from human plasma samples.

Target proteins depleted:

Item	Use	Manufacturer	Albumin	IgG	IgA	IgM	Transferrin	Fibrinogen	a2-Macroglobulin	a1-Antitrypsin	Haptoglobin	a1-Acid Glycoprotein	Apolipoprotein A-I	Apolipoprotein A-II	Complement C3	LDL (ApoB)	Ceruloplasmin	Complement C4	Complement C1q	IgD	Prealbumin	Plasminogen
Enchant™ albumin depletion kit	1	Pall	✓																			
Albumin and IgG spin columns	1	GE Healthcare	✓	✓																		
Qproteome albumin/IgG depletion kit	1	Qiagen	✓	✓																		
Enchant™ Life Sciences kit multi-protein affinity separation	1	Pall	✓	✓																		
MIXED6 IgY	100	Beckman Coulter	✓	✓	✓	✓	✓	✓														
IgY-7 spin column	100	Beckman Coulter	✓	✓	✓	✓	✓	✓		✓												
IgY-12 spin column	100	Beckman Coulter	✓	✓	✓	✓	✓	✓	✓	✓	✓	✓	✓	✓								
IgY14 Microbeads	100	Genway	✓	✓	✓	✓	✓	✓	✓	✓	✓	✓	✓	✓	✓	✓						
ProteoPrep® 20 plasma immunodepletion kit	100	Sigma	✓	✓	✓	✓	✓	✓	✓	✓	✓	✓	✓	✓	✓	✓	✓	✓	✓	✓	✓	✓
PROTEOMINE	1	Oxford Biomedical Research	Top 95% of most abundant proteins																			

Note: ProteomeLab IgY single-component proteome partitioning spin column kits, containing IgY antibodies directed against single abundant proteins, are available for serum albumin, IgG-Fc, transferrin, fibrinogen, HDL, BSA (bovine serum albumin) and RSA (rat serum albumin).

concentration has been achieved by acetone precipitation at -20 °C (Hudgin *et al.*, 1974), but we have found that 41% of the total protein was lost using this method compared with 24% total protein with a membrane ultrafiltration procedure at 10 °C (Millipore Biomax 5K NMWL membrane 0.5 ml ultrafree filters from Millipore UK Ltd., Watford, UK; De Roos *et al.*, 2008a).

Having established best practice in an earlier study examining sample collection procedures and technical variation found in running 2-D gels (De Roos *et al.*, 2008a), we recently applied these recommendations to a human study investigating the effects of an extended fast (36 hours) in 10 healthy volunteers (NuGO Plasma Plus Study, data not shown). Briefly, samples were collected at one site and distributed on dry ice to three collaborating laboratories – each site undertaking analysis of one sample type. To ensure that variation due to sample handling and processing was minimised, the samples were randomised with one restriction; week-4 and the fasting samples from each volunteer were run together in the 1st and 2nd dimension steps. Plasma samples were depleted using the GE Healthcare Albumin and IgG Removal kit, and concentrated with Millipore Biomax 5K NMWL membrane 0.5 ml ultrafree filters. Albumin and IgG depletion of the plasma samples removed 79% of the proteins before the 2-D gels were run (data not shown). A variety of commercial kits are available to determine protein concentration, but it essential to be consistent in the method employed within the study and to ensure the kit is compatible with the samples; some protein assay methods are sensitive to the extraction buffers used in 2-D gel protocols. We used the 2-D Quant kit, also from GE Healthcare. Once the gels have been run, there is a range of stains currently available, and a choice has to be made as to the one that is most appropriate given the facilities and instrumentation used (e.g. Coomassie versus fluorescent stain). We normally use SYPRO Ruby (Invitrogen, Paisley, UK), compatible with the Pharos FX+ Molecular Imager (BioRad, Bath, UK) and the Propick Spot Picker (Genomic Solutions Ltd., Huntingdon, UK), but in the studies described here Flamingo (BioRad) was used instead.

There are a number of commercial 2-D Gel analysis software packages that allow comparison of the gel images and quantification of differences amongst experimental conditions. In the NuGO Plasma Plus studies, two software packages were used. SameSpots™ from Non-Linear Dynamics (Newcastle-upon-Tyne, UK) aligns the images before grouping and then carries out spot detection and, on the assumption that every gel has the same spots on it, any spot-edit affects all gels. The software uses a statistical approach to show the variance between the groups, and allows the images to be grouped in several ways to consider subject variables such as BMI, age, fat intake, etc. without the need for realignment. In contrast, ProteomWeaver (BioRad) groups the images first, before spot detection, and matches the spots amongst the gels. Each gel can (does) have a different number of spots, and each match-edit needs to be checked on all gels, which becomes an issue in large gel sets (> 50). The strengths of ProteomWeaver lie in the ease with which visual observations can be made and images for presentations produced, but each time the images need to be grouped for a different comparison, the images have to be realigned. In both cases, however, the data can be readily exported to other statistical packages for further analysis, which many people prefer, in order to carry out independent statistical analyses.

Concerns about the quality of information provided by proteomics have been expressed by some commentators and referees (Orchard *et al.*, 2003). Thus, a set of guideline have been established for reporting information from proteomic experiments called MIAPE (Minimum Information About a Proteomics Experiment) (Taylor *et al.*, 2007). The gel electrophoresis (Version 1.4, 10th January, 2008) module identifies the minimum information required to report the use of n-dimensional gel electrophoresis in a proteomics experiment in a manner compliant with the aims as laid out in the 'MIAPE Principles' document. More information, along with the standards required by more selective journals, together with the guidelines can be found at http://www.psidev.info/miape/MIAPE_GE_1_4.pdf. Journals are also beginning to request that data are made available for others to repeat the analysis should they so desire. We recommend that these guidelines should be followed from the proposal writing stage to help design and execute the perfect proteomics study. A table cataloguing the reporting conditions used for each gel can also be found at http://www.psidev.info/miape/ MIAPE-GE_Template_v1.1.xls.

Conclusions

- 2-D gels are highly reproducible and can be used to show even subtle effects in humans.
- 2-D gels may point you in the right direction. However, the feasibility of employing this technique in large studies should be considered carefully. If a good idea of the likely proteins changes are known, other techniques (e.g. Multiple Reaction Monitoring-based techniques and quantitative proteomics using stable isotope-labelled standards), which focus on specific the particular proteins of interest, may be more appropriate.
- Get the best you can from your equipment; know the limitations of your analysis!
- Choose your controls carefully.
- Proteomics is non-specific and broad-spectrum and, within a human study, the technique may produce unexpected evidence of anomalous protein levels, which can give rise to ethical issues not necessarily anticipated in the study-design.

Acknowledgements

This work is funded by the BBSRC, Food Standards Agency (FSA) and The European Nutrigenomics Organisation (NuGO). The authors wish to thank Lynda Olivier, Barry Perry, and the other members of the NuGO proteomics consortium: Baukje de Roos, Susan Duthie, Ruan Elliott, Carolin Heim, Hannelore Daniel, Freek Bouwman and Edwin Mariman for their collaboration and contribution in the development of the proteomics work reviewed in this chapter.

6. Proteomic analysis of plasma, platelets and peripheral blood mononuclear cells in nutritional intervention studies

B. de Roos

University of Aberdeen, Rowett Institute of Nutrition and Health, Greenburn Road, Bucksburn, Aberdeen, United Kingdom; b.deroos@rowett.ac.uk

Introduction

Proteomics is emerging as a valuable tool in nutritional research. Proteome analysis of plasma and blood cells such as platelets and peripheral blood mononuclear cells (PBMC) can identify thousands of proteins that:
- Potentially provide valuable new biomarkers for health.
- Reveal early indications of disease risk.
- Assist in establishing dietary responders from non-responders.
- Enable discovery of mechanisms of action for beneficial food components.

Using existing experimental methods it has often proved difficult to deduce the subtle interactions of dietary components in a variable physiological system like the human body (De Roos and McArdle, 2008). In that respect, proteomics technology offers real advantages for investigating these complex interactions. But, it also brings challenges related to study design, sample preparation, standardisation of procedures, and appropriate methods for statistical analysis. Between 2000 and 2008, the number of reviews discussing the technique almost exceeded the number of research papers published using proteomics (De Roos and McArdle, 2008).

Two-dimensional (2-D) gel electrophoresis is still the most widely utilised approach in proteomics for identification of semi-quantitative changes in individual protein levels in tissues, cells and biofluids. Whilst this method is labour intensive, it yields the physical separation of intact polypeptides, providing information on molecular weight and isoelectric point, parameters that can be used in the identification of proteins. The separation can also provide important information about post-translational modifications of proteins. A disadvantage of the method is the difficulty in visualisation and detection of differential regulation of low abundant, and very hydrophobic, acidic, basic or small (<7 kDa) proteins. Proteins involved in inflammatory pathways, such as the cytokines, are secreted and circulate in the blood at concentrations nine orders of magnitude less than albumin. Thus, 2-D gel electrophoresis may not be the most sensitive tool available to reveal effects of nutritional intervention on inflammatory pathways (De Roos and McArdle, 2008). On the other hand, the combination of 2-D gel electrophoresis and mass spectrometry has already proven to be a valuable tool in elucidating changes in pathways that relate to glucose and fatty acid metabolism as well as those proteins associated with oxidative stress and antioxidant defence mechanisms, and redox status (De Roos *et al.*, 2005; Arbones-Mainar *et al.*, 2007; De Roos and McArdle, 2008).

The introduction of difference gel electrophoresis (DIGE) technology allows direct quantitative comparison of differentially-labelled samples using cyanine fluorescent dyes prior to electrophoresis. This method is more reproducible and accurate, and not limited by distortions due to gel-to-gel variation when absolute protein differences between two or three samples is the primary target. Unfortunately, the method is costly and automation difficult.

Shotgun proteomics (i.e. digestion of the protein mixture and multidimensional chromatographic separation of peptides followed by on-line mass spectrometric peptide detection and sequencing) can be combined with stable isotope labelling to enable the quantification of changes in expression levels of hundreds to thousands of proteins in a single experiment. The most commonly adopted approaches include Isotope-Coded Affinity Tags (ICAT) and isobaric Tag Relative Absolute Protein Quantitation (iTRAQ). Quantification is based on relative changes in the levels of labelled peptides, which are common to a family of proteins with differential regulation/abundance. In this case, quantification experiments will lead to ambiguous or conflicting results (De Roos and McArdle, 2008).

Plasma proteomics: an important tool in the search for new biomarkers

The proteomics approaches discussed above are beginning to be used in human dietary intervention studies. Human plasma or circulating cells such as platelets and PBMCs can be obtained relatively easily from study volunteers. The samples are, therefore, primary targets for proteomics to assess qualitative and quantitative changes in physiologically relevant proteins following a specific dietary intervention. The plasma proteome offers the most accessible opportunity to search for proteins that are biomarkers of chronic diseases, and might be altered by diet.

The Human Plasma Proteome Project (HPPP, www.hupo.org) was launched in 2002, under the umbrella of the Human Proteome Organisation (HuPO), to dissect and analyse the human plasma proteome. One of the major challenges in analysing the human proteome was, and remains, the wide concentration range of proteins in plasma/ serum. The protein complement of plasma represents thousands of different proteins with concentrations differing by a factor of 100,000. Nine proteins including albumin, immunoglobulins, transferrin, fibrinogen and haptoglobulin make up 90% of plasma proteins. Depletion of plasma samples, to remove the more abundant proteins, is a generally accepted as step necessary in plasma proteomics enabling 10- to 20-fold higher levels of the remaining proteins to be applied to 2-D gels (De Roos and McArdle, 2008). Increasingly, immuno-affinity is accepted as the most effective sample preparation process for plasma proteomics studies (Fang and Zhang, 2008). Single-use columns may provide more reproducible depletion of abundant protein, and they are less labour-intensive than columns that need to be regenerated between rounds of use (De Roos *et al.*, 2008a). Most immuno-affinity columns deplete 70-90% of the original protein mass (De Roos *et al.*,

2008a) leaving 150-700 µg of protein for proteomic analysis from 30 µl of plasma.

Thus far, few plasma proteomic analyses have been carried out in dietary intervention studies. We assessed the effects of daily fish oil supplements (for six weeks) on the serum proteome using 2-D, MALDI-MS and LC-MS/MS. Serum levels of apolipoprotein A1, apolipoprotein L1, zinc-α-2-glycoprotein, haptoglobin precursor, α-1-antitrypsin precursor, anti-thrombin III-like protein, serum amyloid P component, and haemopexin were all significantly down-regulated (all $P<0.05$) by fish oil compared with high oleic sunflower oil. The decrease in serum apolipoprotein A1 was also associated with a significant shift towards the larger, more cholesterol-rich, high-density lipoprotein 2 (HDL2) particles. The alteration in serum proteins and HDL size imply that fish oil activates anti-inflammatory and lipid modulating mechanisms believed to impede early onset of coronary heart disease (CHD). These proteins are, therefore, potential diagnostic biomarkers to assess the mechanisms whereby fish oils protect against CHD in humans (De Roos *et al.*, 2008b). Levels of plasma apolipoprotein A1 or its isoforms appear to be good examples of plasma proteins that are sensitive to dietary intervention. Another plasma proteomics study, using 2-D and MALDI-MS, indicated that supplementing healthy individuals with α-tocopherol (vitamin E) significantly increased levels of the different isoforms of plasma apolipoprotein A1, in both a time- and dose-dependent manner (Aldred *et al.*, 2006).

Platelet proteomics: mechanistic markers of platelet function and inflammation

For many years 2-D gel electrophoresis has been used to study platelet biology, as the absence of a nucleus prevents platelets from being studied using some molecular biology techniques. Platelets play a major role in haemostasis, and in common disorders like atherothrombosis and coronary artery disease as well as processes relating to vascular integrity, wound healing, and activation of inflammatory and immune responses (Macaulay *et al.*, 2005). Platelets can be obtained in large quantities from relatively small amounts of blood. In general, 100 ml of blood yields ~2×10^{10} platelets, which produce 16-24 mg of protein (Garcia *et al.*, 2005).

The method used for platelet isolation is critical in the interpretation of results of proteome analysis. The platelet proteome is subject to rapid changes in response to external signals, giving rise to potentially large intra- and inter-individual variation. Any method for platelet isolation should aim to isolate a pure platelet preparation whilst minimising platelet activation. Ideally, platelets should be isolated immediately after blood donation to avoid changes in their physiology and viability. Isolation of the upper third of the platelet-rich plasma avoids contamination from other blood cells such as erythrocytes and leucocytes as well as plasma proteins. Platelet-rich plasma should undergo additional centrifugation steps to minimise potential contamination with plasma proteins present in the fraction, which could influence the outcome of the experiment (De Roos *et al.*, 2008a).

Proteomic analysis of platelets has, thus far, included many different approaches. Initial studies focussed on the global cataloguing of proteins present in resting platelets, highlighting the abundance of signalling and cytoskeletal proteins (O'Neill *et al.*, 2002). Changes in the levels of proteins involved in signalling cascades, regulating platelet activation and aggregation with cytoskeletal reorganisation are well described. However, it is less clear whether, for example, dietary intervention can affect levels of cytoskeletal proteins, and if such changes are functionally relevant for signalling cascades. We recently observed that fish oil supplementation for three weeks caused significant up-regulation of several structural proteins detected in the platelets of healthy volunteers by 2-D gel electrophoresis. Such changes could promote cytoskeletal stability of resting platelet, and decrease the incidence of shape-change in response to external stimuli (B. de Roos, unpublished data). Other studies have started to characterise changes in the platelet proteome in response to direct stimulation or other intervention (Macaulay *et al.*, 2005), allowing the identification of novel platelet signalling proteins and phosphorylation events as well as the role of glycosylation in function. These studies provide new insights into the mechanisms of platelet activation and may provide a basis for the development of therapeutic agents for thrombotic disease.

PBMC proteomics: mechanistic markers of immune function and inflammation

PBMCs consist of lymphocytes and monocytes/macrophages. These cells are used to distinguish certain metabolic or disease states based on lymphocyte numbers or their gene expression profiles. Nutritional intervention will affect both the transcriptome and proteome of these cells. For example, intervention with dietary flaxseed differentially regulated 16 PBMC proteins in healthy men (Fuchs *et al.*, 2007a) whereas intervention with an isoflavone differentially regulated 29 PBMC proteins in postmenopausal women (Fuchs *et al.*, 2007b). The latter study suggests the PBMC proteome represents a more sensitive approach for detection of inhibition of inflammatory processes because it responds earlier than classical plasma markers (Fuchs *et al.*, 2007b).

There are a variety of methods for isolating PBMCs from whole blood. However, a significant problem with each of these is contamination of PBMCs with platelets. This contamination impairs the usefulness and validity of PBMC proteomics. Comparison of a range of methods showed the OptiPrep™ method (Sigma Chemical Company, Poole, Dorset, UK) gave the lowest levels of platelet contamination (1:0.8) but the protein yield was correspondingly lower with this method too: 20 ml of whole blood provided on average 15.2×10^6 cells, resulting in approximately 240 µg of protein (De Roos *et al.*, 2008a). Increasing relative platelet numbers (i.e. PBMC contaminated with platelets at a ratio of 1:3 or 1:100) revealed significant changes in overall PBMC protein composition, which was visible on 2-D protein gels. This confirms that platelet contamination can profoundly affect the 2-D protein map of PBMC preparations, and is a serious analytical problem (De Roos *et al.*, 2008a). However, several proteins including α-tropomyosin, fibrinogen and coagulation factor XIII A have been identified as useful biomarkers of platelet contamination for future studies.

Statistical issues affecting data interpretation

Several papers have described the pitfalls of proteomic studies related to experimental design, the misuse of statistical tools available with software analysis packages, and the high rates of 'false positives' in protein identification (Carr *et al.*, 2004; Hunt *et al.*, 2005; Wilkins *et al.*, 2006; Biron *et al.*, 2006; Wheelock and Goto, 2006). A primary challenge for proteomics and other nutrigenomics studies is that the total number of measured proteins is typically larger than the number of samples. Thus, because of multiple hypotheses testing, *P*-values must be considered carefully if the potential for high numbers of false positives is to be avoided. Approaches to adjust *P*-values include reduction of *P*-values to at least 0.01, classical Bonferroni correction, and false discovery rate and q-value methods (Van der Greef *et al.*, 2007).

When using univariate statistical tests such as the Student's t-test, transformation of the data is often necessary because proteomic data are typically not normally distributed but skewed (Wilkins *et al.*, 2006). In general, a good experimental design considers the impact of different sources of variation. Analytical variation, originating from protein separation, staining, and image acquisition and processing steps (De Roos *et al.*, 2008a) can be controlled and minimised to some extent by standardising operating procedures. However, biological variation, originating from environmental or genetic factors as well as differences in sample collection (e.g. in human intervention studies) cannot easily be controlled. In these cases, it is important to process sufficient replicates to ensure adequate statistical power to detect physiological changes in protein levels (Figure 6.1).

Standardisation of proteomics procedures

Proteomics relies on well-developed and validated methods for preparation of plasma and blood cells and subsequent separation and identification procedures (Figure 6.1). Several initiatives are underway to promote standardisation of these procedures. We recently published a study that was designed to develop, optimise and validate protocols

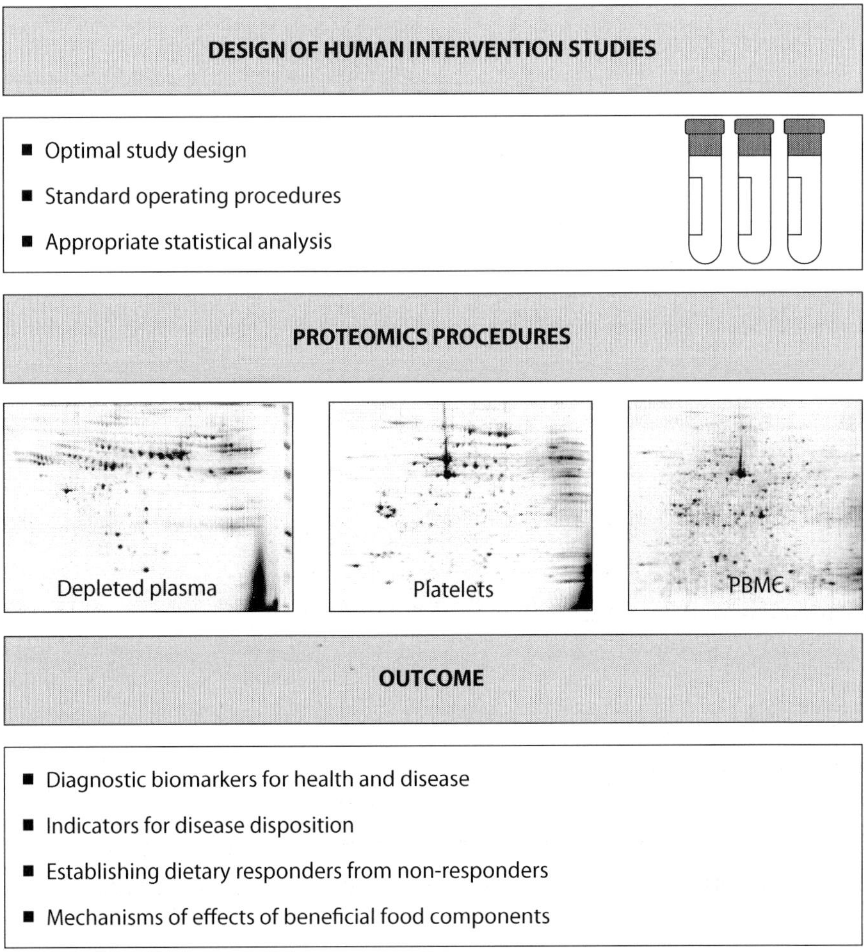

Figure 6.1. Overview of important issues relating to proteomic analysis of plasma, platelets and peripheral blood mononuclear cells in nutritional intervention studies.

for blood processing prior to proteomic analysis of plasma, platelets and PBMCs from human volunteers. Analytical variation of proteins from single samples of depleted plasma, platelet and PBMCs were assessed within and between four laboratories, each using their own standard operating protocols for 2-D gel electrophoresis (De Roos *et al.*, 2008a). However, the extent of inter-individual variability of the human proteome remains unclear (Gerszten and Wang, 2008). This issue is currently being addressed in a Proof-of-Principle Study (NuGO

Design of human nutrigenomics studies

PPS), funded by The European Nutrigenomics Organisation (NuGO), a Network of Excellence funded by the European Commission. The NuGO PPS started in 2007, and aims to determine the biological variability in plasma, platelet, PBMCs, urine and saliva proteomics, plasma and urine metabolomics, and PBMCs transcriptomics. The parameters will be determined at baseline and following a 36-hour fast, and it is hoped this study will not only provide a highly comprehensive human dataset but also insight into both intra- and inter-individual variability at baseline and following metabolic challenge.

Conclusion

The emergence of proteomics platforms, which provide high-resolution protein separation, and mass-spectrometric detection of proteins offer an insight into the complexity of the healthy or diseased phenotype in a high-throughput fashion. Proteomics has an advantage over cDNA micro-arrays, namely that it measures the functional product (protein) of gene expression and allows the identification of protein modifications, which may relate to the activation or inactivation of proteins. Such proteins may have a major physiological role in target organs but might also reflect changes in the mechanisms of action initiated by dietary intervention. A major advantage of the proteomics platform is the qualitative and quantitative analysis of human body fluids, which can be obtained relatively easily from components of donated blood including plasma, serum and circulating cells, such as platelets and PBMCs. Proteomics has the potential to deliver specific and relevant diagnostic biomarkers for health and reveal early indications of disease susceptibility. Considering the requirements for new diagnostic biomarkers in clinical practice, it is envisaged that such indicators will be obtained, in the main, from (depleted) plasma. New biomarkers obtained from platelets and PBMCs, on the other hand, could be used to provide valuable insight into changes in the mechanisms associated with, for example, inflammation, haemostasis, and immune function.

Acknowledgements

Work in the author's laboratory is funded by the Scottish Government Rural and Environment Research and Analysis Directorate (RERAD). The author wishes to thank the members of the NuGO proteomics consortium: Susan Duthie, Abigael Polley, Francis Mulholland, Ruan Elliott, Ian Johnson, Carolin Heim, Hannelore Daniel, Freek Bouwman and Edwin Mariman for their collaboration and contribution in the development of the proteomics work reviewed in this chapter.

7. Human nutrigenetic research: the strengths and limitations of various study designs

A.M. Minihane

Hugh Sinclair Human Nutrition Group, School of Chemistry, Food Biosciences and Pharmacy, University of Reading, Reading, United Kingdom

Introduction

Nutrigenetics is the branch of nutrition concerned with the interaction between genotype and dietary components to influence metabolism, health status and risk of diet-related diseases. These interactions are now recognised to be highly complex with the influence of genetic make up on metabolic homeostasis and health known to be affected by numerous behavioural components, including diet. Likewise, an individual's genetic make-up can influence food intake and the impact of altered dietary composition on physiological processes, with genotype influencing digestion, absorption, post-absorptive metabolism and cellular responsiveness to dietary components.

Numerous different types of genetic variation exist with monogenic Mendelian disorders, such as Huntington's disease, familial hypercholesterolaemia and cystic fibrosis, with development attributable to significant changes in individual gene structures including repeat sequences and insertions or deletions of relatively large sections of DNA. In contrast, risk of chronic polygenic disorders such as cardiovascular disease (CVD), obesity, cancers, etc. is thought to be determined in part by the cumulative effects of changes in individual nucleotides (A, T, C or G) in the DNA sequence referred to as single nucleotide polymorphisms (SNPs).

In isolation SNPs are not causative, but predictive to varying degrees of the likelihood that someone will develop a particular illness. SNPs exert their metabolic effects by altering either the concentration of the protein produced or its amino acid sequence, either of which may influence its metabolic functionality. In addition to individual SNPs, many studies are also looking at haplotypes; a group of SNPs that are in the same chromosome region, which are statistical associated and inherited together. As such, the identification of several SNPs within a haplotype block can be used as a biomarker for all other polymorphic sites in the same region. The International HapMap Consortium (www.hapmap.org) has advanced greatly the knowledge-base in this area, with their website providing detailed information on the origins, content and usability of this resource.

Currently, there is a large amount of research activity underway that is attempting to identify SNPs/haplotypes, which are associated with disease risk and respond to diet or other behavioural attributes. A wide range of research approaches, each with different strengths and limitations, are being used. The focus of this chapter is on the approaches used in human volunteers. The ultimate aim of this work is – in the future – to use genetic profiling as a means for early detection of disease risk and the personalisation of dietary recommendations for individuals or population subgroups (Figure 7.1). It is likely that such a personalised approach, along with increasing consumer motivation to adopt lifestyle changes, will increase the physiological benefit afforded to the individual(s). Although it is recognised that epigenetics (DNA methylation status) is also emerging as a genetic determinant of disease risk and responsiveness to diet, this area topic is discussed in Chapter 8.

Basic components of all nutrigenetics studies

Human nutrigenetics studies can be broadly categorised into epidemiological studies or intervention trials, and three basic components are measured or controlled by the researcher:
- One or more physiological outcome(s), which typically are biomarkers, surrogate end points and/or disease occurrence. For example, in cardiovascular disease, these would be blood pressure

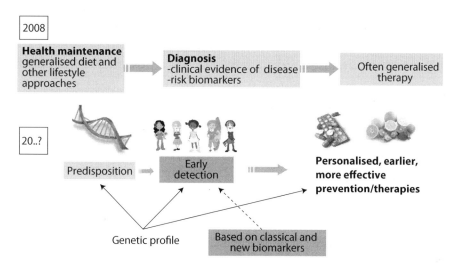

Figure 7.1. Schematic of the potential for genetic, nutrigenetics and pharmacogenetics in health maintenance and treatment of diseases.

(biomarker), carotid-intima-media thickness (surrogate end point) and myocardial infarction (disease occurrence).

• Dietary assessment or manipulation of one or more dietary components.

• Genotype.

Epidemiological studies

Applying candidate gene approaches in case-control or cohort epidemiological studies has to date provided the vast majority of information on genetic-diet-disease risk associations. These studies are hypothesis-driven with the SNPs/haplotypes chosen on the basis of their known role in the phenotype or metabolism of the dietary component of interest. In these studies, dietary composition is surveyed using a variety of dietary assessment techniques such as food frequency questionnaires, 24-hour recall or weighed- or estimated dietary diaries. A candidate gene approach has been used successfully in the Framingham Heart Study to identify a number of SNPs the penetrance of which with respect to blood lipid profile is dependent on the dietary fat composition (Ordovas, 2002; Ordovas *et al.*, 2002). Another highly cited example, identified using a candidate gene approach, is the association between the MTHFR (methylenetetrahydrofolate reductase) SNP, folate intake, plasma homocysteine and CVD risk, with the penetrance of the genotype reported (although not consistently) to be lower in those with adequate folate status (Klerk *et al.*, 2002).

Over the last 10 years, the use of a non-hypothesis-driven alternative to the candidate-gene approach, namely genome wide analysis (GWA) in large cohorts of unrelated individuals, such as the Wellcome Case-Control Consortium (WCCC), has dramatically increased the number of genetic discoveries (Pearson and Manolio, 2008). In these studies, genetic variation in up to 80% of the total DNA is assessed typically using between 100,000-1,000,000 marker SNPs and information from the HapMap project. Up to 100 susceptibility gene loci for a range of complex polygenic common chronic diseases have been confirmed, identified and replicated using these GWA approaches (Consortium WTCC, 2007). However, they have contributed little to our understanding of diet-genotype interactions as most of these studies have not captured or reported any dietary information. Future applications of GWA to cohorts such as Framingham, EPIC, Nurses Health Study and the Health Professionals Study, where dietary data are available, will undoubtedly advance current nutrigenetics knowledge.

A current concern, when considering the literature based on both the candidate-gene and GWA approaches, is that many reported associations fail to be consistently replicated in independent studies. The reasons for these inter-study inconsistencies are likely to be many. Some studies may be under-powered, which can lead to a failure to detect the subtle physiological impact of a particular gene variant and provide an imprecise estimate of the size effect. Furthermore, emerging evidence indicates that an array of physiological attributes impact on the penetrance of a particular gene variant and that the apparent inconsistencies represent a change in the size of the genotype effect under a particular set of circumstances. Perhaps the most widely researched, and consistently observed, genotype-phenotype association is that between the apoE4 variant and CVD risk (Song *et al.*, 2004). However, there is good evidence to suggest that the effect of the apoE

Design of human nutrigenomics studies

genotype is both sex- and age-dependent with the greatest impact in young males. For example, in the Helsinki Sudden Death Study, which conducted lesion staining in the coronary arteries of 700 individuals, significant age-genotype interactions were observed with a significant impact of genotype only in the under 53 years-of-age group (Ilveskoski *et al.*, 1999). The reason for this reduction in size effect of genotype with age is probably because an individual accumulates environmental influences (Zdravkovic *et al.*, 2002) with age and develops a multi-faceted higher-risk phenotype, which may mask the relatively modest physiological impact of individual genotypes.

In a recent and aptly named publication 'On the replication of genetic associations: timing can be everything' the authors highlight the importance of 'age-varying associations' in the interpretation of results from observational trials (Lasky-Su *et al.*, 2008). It is likely that age and other physiological parameters such as sex, disease status, body weight etc. should be considered when designing and comparing data from human intervention nutrigenetics trials.

Undoubtedly the greatest sources of inconsistencies when conducting and comparing observational nutrigenetics studies are the accuracy of the dietary information provided by the participants, and the accuracy and completeness of nutrient composition tables used to translate dietary information into intake of individual dietary components. Inaccuracy in this information is particularly problematic for some population groups such as obese individuals or those with diagnosed disease either of whom may have an over-inflated view of the 'healthiness' of their diet. It is particularly pertinent to consider the impact of disease diagnosis on perception of diet when conducting case-control studies. The availability, accuracy and completeness of food composition tables are a specific issue for a number of micronutrients and non-nutrient dietary components. The identification of biomarkers for intake in accessible biofluids such as blood, urine and saliva is currently being investigated as an alternative to traditional dietary assessment techniques. To date, however, such quantitative biomarkers are currently unavailable for most dietary components. The provision of foods or dietary components in the form of a supplement or food extract overcomes this inability to assess accurately dietary exposure in observational studies.

Intervention nutrigenetics trial

In nutrigenetic intervention trials, genotyping can be done prospectively or retrospectively, or both. In prospective genotyping, volunteers are recruited to the study according to their genotype with equal numbers in each group. In retrospective genotyping, the genetic profiling is more opportunistic and conducted once the study is complete. In the limited number of prospective genotyping studies conducted to date, retrospective genotyping for other SNPs is often undertaken. In nutrigenetics human trials, the same issues need to be considered when designing the study as might be thought about if the study did not have a genotyping component. In demonstrating this, the text will refer to a number of nutrigenetics studies completed in our laboratory as examples (Olano-Martin *et al.*, unpublished data; Minihane *et al.*, 2000; Caslake *et al.*, 2008). These studies have examined the impact of apoE genotype on the blood lipid response to the fish oil fatty acids eicosapentaenoic acid (EPA) and docosahexaenoic acid (DHA).

When designing a nutrigenetics intervention trial, consideration should be given to:
- Clearly defined hypothesis and research questions.
- Choice of placebo treatment.
- Length of intervention period – in chronic studies the period intervention will largely be determined by the likely responsiveness of the disease marker of interest. For the blood lipid response to EPA and DHA, a minimum of four weeks was required for the blood lipid effects to begin to plateau.
- Study design – cross-over versus parallel. A cross-over study is always desirable to minimise the heterogeneity between control and treatment interventions. For a long intervention period (e.g. several months), however, a parallel design may be preferable for logistical reasons as well as any concerns about participants compliance rates, which may suffer with two or more long intervention periods in a cross-over design.

- Inclusion and exclusion criteria – examples for one of the above mentioned apoE trials are given in Table 7.1. When deciding on these criteria it is important to consider for whom the results are applicable. Specifically, in nutrigenetics studies it is important to realise that inclusion/exclusion criteria may bias against a particular genotype subgroup. For example, as an apoE4 genotype is associated with a modest increase in total cholesterol having a cut-off of 8.0 mmol/l (Table 7.1) may exclude those most sensitive to the effects of the genotype of interest.
- In prospective genotyping studies, the groups need to be matched at baseline for important determinants of response to treatment such as age, sex etc. However, matching genotype groups at baseline for factors affected by the genotype under examination should be avoided as this will introduce bias into the design. For example, in the

Table 7.1. Inclusion and exclusion criteria for the FINGEN study (Caslake et al., 2008).

Age groups 20-70 years

BMI 18.5-30 kg/m^2

Total cholesterol <8.0 mmol/l, TAG <3.0 mmol/l and glucose <6.8 mmol/l

Not suffered a myocardial infarction in the previous two years

Not on drug treatment for hyperlipidaemia or inflammatory conditions or regular (daily) aspirin users

Not suffering from diabetes or other endocrine disorders

Not taking any dietary supplements including fatty acids (e.g. fish oils, evening primrose oil) or prepared to stop taking supplements prior to the trial

Not consuming high doses of antioxidant vitamins (A, C, E, β-carotene) – maimum permitted intake is 800 µg vitamin A, 60 mg vitamin C, 10 mg vitamin E and 400 µg β-carotene, respectively, equivalent to daily dose present in a standard multivitamin capsule

Not consuming more than one serving of oily fish per week

Not vigorous exercisers (>3 ´ 30 minute vigorous sessions per week)

Not planning to lose weight

Not pregnant, lactating or planning a pregnancy

Not suffering from any life-threatening illness

---

apoE genotype fish oil studies, the groups were matched at baseline for age, sex and BMI, but not blood lipid profiles.

- Sample size – for retrospective genotyping studies, this is a particular concern. A small sample size in the rare allele subgroup can often lead to the conclusion that there is no genotype-diet interaction when an interaction does in fact exist, but was not observed because of the small sample size. This often leads to the reporting of false negatives in the literature. For example, in an intervention group of n=50, examination of the impact of being homozygous allele A for a particular gene, which is only present in 10% of the population, would be conducted by comparing the responsiveness of 45 people against only five individuals. Unless the genotype-diet interaction was very large and consistent, such a comparison would fail to reach statistical significance. For retrospective genotyping large groups of several hundred participants are needed, but they can only be used to test the impact of genotypes with greater than 10% rare allele frequency.
- Capture of as much participants' information as possible since it is likely that a large range of physiological and behavioural factors will impact the genotype-diet-phenotype associations, all of which should be included in the statistical model(s).

Ethical considerations for human nutrigenetics studies

Ethical consent needs to be obtained from all volunteers before any genotyping is conducted. It is also important that all investigators have sufficient knowledge of the principles of genetics to allow them to provide volunteers with information about the meaning of specific genotypes. Consideration should also be given as to whether this information is made available to the volunteers or not, and ethics committees can take a different stance on this. Based on the experience of investigators working on nutrigenetics studies, it is evident that volunteers place much more emphasis on being carriers of an 'at-risk' allele compared with far more significant disease risk factors such as smoking or excess body weight. From an ethical perspective, it is important therefore that the risks associated with individual genotypes are explained carefully to the volunteer, if genotype information is to be released. For a more comprehensive description of bioethics in nutrigenetics, please refer to the NuGO online Bioethics Tool (nugo. dife.de/bot/).

Genotyping techniques

Relative to other nutrigenomics technologies such as transcriptomics, proteomics or metabolomics, genetic profiling is a more robust process with sampling techniques and post-sampling processing unlikely to impact on the genetic profile. DNA is extracted from the human tissues (e.g. biopsy, hair, bucal swab, buffy coat) or biofluids (e.g. whole blood or saliva) using traditional multi-step isolation techniques or, more typically, using commercially available DNA isolation kits or instruments. When conducting a human intervention trial where blood samples are collected, DNA is often extraction from the buffy coat (i.e. white cell layer). More than 100 techniques are currently available to genotype isolated DNA with the choice determined by factors such as available equipment, the number of SNPs being assayed, and the degree of high-throughput required. Techniques such as DNA fragment analysis, PCR, MALDI-TOF, mass spectrometry and sequencing are typically used when the samples are analysed for a relatively low number of SNPs. For GWA studies, SNP arrays that assay for 100,000+ SNP are used. An increasing number of organisations that conduct SNP analysis commercially, in a cost cost-effective manner, are also available.

Conclusion

Common gene variants may determine disease risk and response to dietary change, and there is currently a large amount of research ongoing in this area. However, data derived from individual nutrigenetic studies are often inconsistent due to poor study design, and data collection and analysis. For observational trials, inaccurate assessment of diet composition is hugely problematic, and the development of improved dietary assessment techniques along with suitable biomarkers is needed. For human intervention trials with retrospective genotyping, studies often lack the power necessary for conducting meaningful genotype-diet interaction analysis. There needs to be a greater appreciation of this, and the fact that it leads to false-negative results. A human intervention trial, with prospective genotyping, and sufficient power as well as equal numbers in each genotype group can overcome these difficulties. But, cost and logistical difficulties mean such an approach cannot be adopted to test large numbers of potential genotype-diet-phenotype interactions, and should instead be reserved for interactions that are of wide public health interest and/or have been repeatedly observed in observational studies.

8. Nutritional epigenomics in human studies

J.C. Mathers

Human Nutrition Research Centre, Institute for Ageing and Health, Newcastle University, William Leech Building, Framlington Place, Newcastle-upon-Tyne, United Kingdom; john.mathers@ncl.ac.uk

Epigenetics describes genomic processes, often leading to changes in transcription, which are heritable from one cell generation to the next without involving changes to the DNA sequence. The primary epigenetic marks include changes to histone 'decoration' (i.e. covalent addition of methyl, acetyl, phosphate and ubiquitin groups) and DNA methylation. The totality of epigenetic marks in a given genome under particular circumstances is known as the epigenome. Cytosine residues, normally when followed by a guanine (a CpG dinucleotide), are the focus for DNA methylation marks and the majority of CpGs in the genome are methylated in healthy people. In contrast, unusually dense clusters of CpGs (known as CpG islands) in the promoter regions of genes are normally unmethylated, and methylation of these domains results in gene silencing. The pattern of epigenetic marks on the genome is modified by a wide range of environmental exposures including diet and, because of their key role in regulating transcription, it has been hypothesised that epigenetic mechanisms are responsible for a significant portion of the links between nutrition and health.

DNA methylation marks are complex displaying considerable variability between CpG sites within genomic regions and amongst tissues. Fortunately, the technologies to facilitate DNA methylation studies in humans are improving rapidly, and the wider availability and lower cost are making them accessible to nutrition researchers. To date, there have been few nutritional epigenomics studies in humans but there is proof-of-principle that the supply of certain nutrients (e.g. folate) alters genomic DNA methylation in adult humans. The design issues specific to such studies are poorly understood but, as with other omics studies in humans, it is probable that the significant design issues will include: (1) optimisation of protocols to cope with intra- and inter-individual variation; (2) use of surrogate tissues where access to the tissue of interest is difficult or impossible for practical or ethical reasons; and (3) identification of significant confounding factors and the development of strategies to minimise their effects.

Epigenetics and epigenomics

The ample evidence that dietary exposure is a major modifier of health at all stages of the life-course has stimulated research on the molecular mechanisms that are responsible for mediating these effects. In addition to providing energy for fuelling cellular processes and substrates for the synthesis of cellular constitutes, the most important way in which food components are likely to influence health is through altered gene expression. Although post-transcriptional regulation of gene expression is known, most of the mechanisms for temporal and tissue-specific regulation of gene expression act at the level of transcription initiation (for overview of nutritional modulation of gene expression, see Mathers, 2006a). More recently, it has become apparent that epigenetic mechanisms may also be important as a mechanistic link between dietary exposure and gene expression with significant implications for cell function and health.

Epigenetics describes genomic processes, often leading to changes in transcription, which are heritable from one cell generation to the next without involving changes to the DNA sequence. The primary epigenetic marks include: (1) changes to the constellation of groups (i.e. methyl, acetyl, phosphate and ubiquitin covalently bound to amino acid residues), which decorate the tails of histones around which DNA is wrapped in nucleosomes; and (2) DNA methylation. The vast majority of methyl group additions to DNA occur at the 5' position in cytosine residues where the cytosine is followed by a guanine (a CpG dinucleotide). Overall, CpG dinucleotides are under-represented in the mammalian genome, but dense assemblies of CpGs (known as CpG islands, CGI) are found throughout the genome most notably in the promoter regions of about half of all human genes. In healthy young tissues, approximately 70% of CpGs are methylated but the cytosines in the CGI promoter regions are usually unmethylated. Whilst the functional consequences (if any) of many individual histone modifications are still to be discovered, it is clear that genomic regions with substantial acetylation of histones (e.g. at lysine residues at positions 9, 14, 18 and 27 in histone H3) (Bernstein *et al.*, 2007) are associated with expression of genes in the vicinity whereas methylation of the lysine residue at position 4 in histone H3 usually signals transcriptional silencing (Hake *et al.*, 2004). Active transcription may be expected for genes with unmethylated CGI in their promoters whilst methylation acts to 'switch off' expression.

In most cases, the identities of the specific CpG residues whose methylation status is critical in determining whether a gene is switched on or switched off are unknown. However, a reasonable case can be made that gene silencing by promoter methylation may occur at DNA domains involved in transcription factor binding (Mckay *et al.*, 2009 in press), and it is probable that DNA methylation and chromatin changes act in concert to regulate gene expression (Bernstein *et al.*, 2007). The totality of epigenetic marks in a given genome under particular circumstances is known as the epigenome. To date, there have been few studies of the interactions between diet and the epigenome in humans, and this paper will review only some of the underlying science (focusing on DNA methylation) on which human studies can be developed.

Epigenetics and phenotypic plasticity

Many of DNA methylation marks are established in embryonic and fetal life where the genome undergoes radical changes in epigenetic marks, characterised by waves of de-methylation of DNA followed by re-methylation, during the first few cell divisions (Reik *et al.*, 2001). Different cell lineages (giving rise to different organs and tissues) have distinctive methylation patterns as part of the differentiation process. However, DNA methylation marks are not fixed for life, and undergo changes in response to a range of environmental exposures and stochastic events. Because they are copied from one cell generation to the next, epigenetic marks are attractive candidates for propagating cellular memory of environmental exposure. In this way, the pattern of epigenetic marks might provide a molecular mechanism for underpinning the Developmental Origins of Adult Health and Disease hypothesis, which suggests early life exposures in humans are a major determinant of health, decades later. This leads to the hypothesis that epigenetics may be a key mechanism linking environmental (food) exposure with gene expression, and so enable phenotypic plasticity in the context of a fixed genotype. Evidence in support of this hypothesis comes from studies of a wide range of pre-natal and early post-natal exposures including dietary components, xenobiotic chemicals, behavioural cues and low-dose radiation, which leave their mark on the epigenome and influence disease risk in later life (Jirtle and Skinner, 2007).

Factors affecting DNA methylation in humans

The B vitamin folate is important in DNA methylation because it is required for the synthesis of S-adenosyl methionine (SAM), the universal methyl donor that provides methyl groups for DNA methylation reactions catalysed by DNA methyl transferases (DNMTs). With limiting concentrations of SAM, as might be expected when folate supply is sub-optimal, there is likely to be competition between methyltransferases for the available SAM, and changes in DNA methylation would be anticipated.

Effects of reduced folate supply were tested in healthy, non-smoking post-menopausal women who resided in a metabolic unit for 91 days (Jacob et al., 1998). When folate intake was reduced from 195 μg/day (sufficient to meet the needs of non-pregnant, non-lactating adults) to 56 μg/day, the expected fall in plasma folate concentration and corresponding rise in plasma homocysteine (tHcy) were observed (Jacob et al., 1998). Raising folate intake in steps up to 516 μg/day (using folic acid supplements) restored plasma folate concentration, but changes in plasma tHcy were less marked. Genomic methylation in DNA from lymphocytes (estimated using a methyl acceptance assay) responded to the altered folate status as predicted, decreasing when folate intake was reduced and returning towards baseline levels following the highest intake of folate (diet plus folic acid supplements,

Jacob *et al.*, 1998). These data suggest that genomic DNA methylation levels are plastic and respond to folate supply over quite short time periods (a few days). Nijhout *et al.* (2006) used a mathematical model of the methionine cycle (from which SAM is produced) to predict the consequences of altering supplies of folate and methionine on DNMT reaction rate. A 50% restriction in folate supply produced a modest fall in DNMT reaction rate, but DNMT reaction rates fell off rapidly when this degree of folate restriction was coupled with reduced methionine supply (Nijhout *et al.*, 2006). The genomic location of the cytosine residues whose methylation status was altered in response to changes in dietary folate intake in the study by Jacob *et al.* (1998) is unknown, but might include non-coding regions of the genome, for example, DNA repetitive elements such as Alu elements and long interspersed nucleotide elements (LINE), which are normally heavily methylated (Yang *et al.*, 2004). Whilst hypomethylation of these elements may have adverse consequences (e.g. may result in activation of retrotransposons and induction of chromosomal instability, Wilson *et al.*, 2007), the functional sequelae are poorly understood.

From a health perspective, changes in the methylation status of CGI in the promoter regions of genes, which are causally related to altered transcription, are more obvious candidates for further study. To date there is limited understanding of characteristics of genes or genomic sequences that are susceptible to altered regulation of expression due to changes in DNA methylation. But, Mckay *et al.* (2009 in press) outlined a bioinformatics strategy for identifying such genes starting from whole genome expression data. As an alternative or complementary approach, Nijhout *et al.* (2006) proposed that the availability of kinetic data on the mechanisms that expose CGI to methylation would allow their model to be extended to include effects of altered methionine and folate supply on differential methylation of different genes.

In addition to modulation in response to altered methyl donor supply, patterns of DNA methylation also change with age. In many cases, older age is accompanied by an increase in the methylation of promoter regions of genes, which show little (if any) methylation in younger individuals. Since these age-related methylation changes are observed in genes with key functional roles (e.g. cell cycle regulatory genes, genes involved in WNT signalling and tumour suppressor genes, Belshaw *et al.*, 2008), a case can be made that such changes may contribute to the reduced function and altered phenotype that characterise ageing (Mathers, 2006b). Some of the best evidence for environmental determinants of changes in DNA methylation with age comes from the study of monozygotic twins by Fraga *et al.* (2005). This study showed that the DNA methylation patterns of younger twin pairs were much more similar than those of older twin pairs, despite identical genotypes and more importantly that these divergent DNA methylation patterns were accompanied by less similar gene expression portraits (Fraga *et al.*, 2005). The impact of environmental influences, in addition to those of age, is illustrated by the observation that inter-twin differences in DNA methylation patterns were magnified if the twins lived apart (Fraga *et al.*, 2005).

Variation in DNA methylation marks

Although variation in DNA methylation marks is a key component of the emerging human epigenome project (Brena *et al.*, 2006), systematic study of DNA methylation marks at gene locus, cell type, tissue and individual levels is in its infancy. However, it is apparent that there is considerable variation at each of these levels of organisation. For example, in a proof-of-principle study Rakyan *et al.* (2004) showed that each of 27 CpGs in the CYP21A2 gene and 13 CpGs in the TNF gene were methylated to different extents, and that there was extensive variation at these loci amongst tissues and individuals. Evidence of the richness and complexity of methylation marks has been provided by 'conventional' bisulphite sequencing of human chromosomes 6, 20 and 22, which obtained 2×10^6 CpG methylation values from 12 tissues (Eckhardt *et al.*, 2006). Today's state-of-the-art approach for large scale methylation analysis is array-based, but technological developments that facilitate much higher throughput analysis (e.g. massively parallel bisulphite pyrosequencing) are now becoming available (Korshunova *et al.*, 2008). The functional consequences of these different methylation patterns including, for example, the identity of specific CpGs, which are critical for regulating gene expression, remain to be established.

At a whole person level, variation in DNA methylation marks could arise from the impact of individual genotypes. For example, having the TT version of the most common (677C→T) polymorphism in the MTHFR gene is associated with a significant reduction in genomic DNA methylation (32 versus 62 ng 5-methylcytosine/μg DNA for TT and CC polymorphisms, respectively, Friso *et al.*, 2002). That study also showed that whilst DNA methylation correlated positively with folate status and negatively with plasma tHcy, the effect of low folate status on DNA methylation was confined to those with the TT genotype (i.e. providing evidence for a nutrition-genotype interaction, Friso *et al.*, 2002). The genes encoding DNMTs are an obvious area of interest for genetic influences on DNA methylation. There are at least five members of the DNMT gene family with DNMT1 being the house-keeping gene whose protein product is responsible for maintaining DNA methylation patterns during mitosis (Bestor, 2000). A number of variants in DNMTs have been described and, in a case-control study of approximately 9,000 individuals, a variant in DNMT3b (a *de novo* methyltransferase) was found to correlate with breast cancer risk but no effect on global DNA methylation or on the extent of methylation of the promoters of 15 cancer-related genes was detected (Cebrian *et al.*, 2006).

Towards nutritional epigenomics studies in humans

The National Institutes of Health (2008) in the USA have identified epigenetics as an emerging frontier of science. There is likely to be a substantial increase in resources focussed on epigenetics research at all levels from developments of new high-throughput technologies for assessing methylation marks across the genome and application with large numbers of samples (e.g. in epidemiological studies) to bioinformatics approaches for interpreting the burgeoning data, which will emerge from such studies. These developments will provide important new tools for nutrition researchers investigating how dietary exposures or nutritional status influences the formation and maintenance of DNA methylation marks, and the role these epigenetic patterns play in modulating health throughout the life-course. The resulting research may help to answer questions such as (1) which genes, contributing to altered disease susceptibility, are epigenetically deregulated by dietary factors; (2) which dietary factors, in what doses, affect epigenetic markings; (3) in what periods of life is the genome especially vulnerable to altered epigenetic markings by nutrition; and (4) how quickly do epigenetic marks respond to dietary interventions, and are these responses sustained.

In summary, epigenetic marks (DNA methylation) are an important mechanism linking environmental exposures including nutritional exposures with gene expression and cell function, leading to altered phenotype in the context of fixed genotype. In addition, genetic factors may interact with nutrition to modify CpG methylation. DNA methylation marks are complex showing considerable variability between CpG sites within genomic regions and amongst tissues. Fortunately, the technologies to facilitate DNA methylation (and other components of epigenomics) studies in humans are improving rapidly, and the wider availability and lower cost are making them accessible to nutrition researchers. To date, there are few examples of nutritional epigenomics studies in humans, and the design issues specific to such studies are poorly understood. As with other omics studies in humans, however, it is probable that these issues will include: (1) optimisation of protocols to cope with intra- and inter-individual variation; (2) use of surrogate tissues where access to the tissue of interest is difficult or impossible for practical or ethical reasons; and (3) identification of significant confounding factors and development of strategies to minimise their effects.

Acknowledgements

Research on nutritional epigenomics in my laboratory has been funded by the Biotechnology and Biological Sciences Research Council through the Centre for Integrated Systems Biology of Ageing and Nutrition (BB/C008200/1), the Food Standards Agency (N12004), the World Cancer Research Fund (2001/37) and by NuGO 'The European Nutrigenomics Organisation; linking genomics, nutrition and health research' (NuGO; CT-2004-505944) which is a Network of Excellence funded by the European Commission's Research Directorate General

Design of human nutrigenomics studies

under Priority Thematic Area 5, Food Quality and Safety Priority, of the Sixth Framework Programme for Research and Technological Development. Further information about NuGO and its activities can be found at http://www.nugo.org.

9. References

Afman, L. and Muller, M., 2006. Nutrigenomics: from molecular nutrition to prevention of disease. Journal of the American Dietetic Association 106(4): 569-576.

Aldred, S., Sozzi, T., Mudway, I., Grant, M.M., Neubert, H., Kelly, F.J. and Griffiths, H.R., 2006. Alpha tocopherol supplementation elevates plasma apolipoprotein A1 isoforms in normal healthy subjects. Proteomics 6: 1695-1703.

Arbones-Mainar, J.M., Ross, K., Rucklidge, G.J., Reid, M., Duncan, G., Arthur, J.R., Horgan, G.W., Navarro, M.A., Carnicer, R., Arnal, C., Osada, J. and De Roos, B., 2007. Extra virgin olive oils increase hepatic fat accumulation and hepatic antioxidant protein levels in APOE(-/-) mice. Journal of Proteome Research 6: 4041-4054.

Bedair, M. and Sumner, L.W., 2008. Current and emerging mass-spectrometry technologies for metabolomics. Trends in Analytical Chemistry 27(3): 238-250.

Belshaw, N.J., Elliott, G.O., Foxall, R.J., Dainty, J.R., Pal, N., Coupe, A., Garg, D., Bradburn, D.M., Mathers, J.C. and Johnson, I.T., 2008 Profiling CpG island field methylation in both morphologically normal and neoplastic human colonic mucosa. British Journal of Cancer 99: 136-142.

Bergheanu, S.C., Reijmers, T., Zwinderman, A.H., Bobeldijk, I., Ramaker, R., Liem, A.-H., Van der Greef, J., Hankemeier, T. and Jukema, J.W., 2008. Lipidomic approach to evaluate rosuvastatin and atorvastatin at various dosages: investigating differential effects among statins. Current Medical Research and Opinion 24(9): 2477-2487.

Bergmann, M.M., Görman, U. and Mathers, J.C., 2008. Bioethical considerations for human nutrigenomics. Annual Review of Nutrition 28: 447-467.

Bernstein, B.E., Meissner, A. and Lander, E.S., 2007. The mammalian epigenome. Cell 128: 669-681.

Bestor, T.H., 2000 The DNA methyltransferases of mammals. Human Molecular Genetics 9: 2395-2402.

Bijlsma, S., Bobeldijk, I., Verheij, E.R., Ramaker, R., Kochhar, S., Macdonald, I.A., Van Ommen B. and Smilde, A.K., 2006. Large-scale human metabolomics studies: a strategy for data (pre-) processing and validation: Analytical Chemistry 78(2): 567-574.

Biron, D.G., Brun, C., Lefevre, T., Lebarbenchon, C., Loxdale, H.D., Chevenet, F., Brizard, J.P. and Thomas, F., 2006. The pitfalls of proteomics experiments without the correct use of bioinformatics tools. Proteomics 6: 5577-5596.

Bouwens, M., Afman, L.A. and Müller M., 2007. Fasting induces changes in peripheral blood mononuclear cell gene expression profiles related to increases in fatty acid beta-oxidation: functional role of peroxisome proliferator activated receptor alpha in human peripheral blood mononuclear cells. American Journal of Clinical Nutrition 86(5): 1515-1523.

Bouwens, M., Afman, L.A. and Müller, M., 2008. Activation of peroxisome proliferator-activated receptor alpha in human peripheral blood mononuclear cells reveals an individual gene expression profile response. BMC Genomics 9: 262.

Brena, R.M., Huang, T.H. and Plass, C., 2006. Toward a human epigenome. Nature Genetics 38: 1359-1360.

Carr, S., Aebersold, R., Baldwin, M., Burlingame, A., Clauser, K. and Nesvizhskii, A., 2004. The need for guidelines in publication of peptide and protein identification data: Working group on publication guidelines for peptide and protein identification data. Molecular and Cellular Proteomics 3: 531-533.

Caslake, M.J., Miles, E.A., Kofler, B.M., Lietz, G., Curtis, P., Armah, C.K., Kimber, A.C., Grew, J.P., Farrell, L., Stannard, J., Napper, F.L., Sala-Vila, A., West, A.L., Mathers, J.C., Packard, C., Williams, C.M., Calder, P.C. and Minihane, A.M., 2008. Effect of sex and genotype on cardiovascular biomarker response to fish oils: the FINGEN study. American Journal of Clinical Nutrition 88(3): 618-629.

Cebrian, A., Pharoah, P.D., Ahmed, S., Ropero, S., Fraga, M.F., Smith, P.L., Conroy, D., Luben, R., Perkins, B., Easton, D.F., Dunning, A.M., Esteller, M. and Ponder, B.A., 2006. Genetic variants in epigenetic genes and breast cancer risk. Carcinogenesis 27(8): 1661-1669.

Cobb, J.P., Mindrinos, M.N., Miller-Graziano, C., Calvano, S.E., Baker, H.V., Xiao, W., Laudanski, K., Brownstein, B.H., Elson, C.M., Hayden, D.L., Herndon, D.N., Lowry, S.F., Maier, R.V., Schoenfeld, D.A., Moldawer, L.L., Davis, R.W., Tompkins, R.G., Baker, H.V., Bankey, P., Billiar, T., Brownstein, B.H., Calvano, S.E., Camp, D., Chaudry, I., Cobb, J.P., Davis, R.W., Elson, C.M., Freeman, B., Gamelli, R., Gibran, N., Harbrecht, B., Hayden, D.L., Heagy, W., Heimbach, D., Herndon, D.N., Horton, J., Hunt, J., Laudanski, K., Lederer, J., Lowry, S.F., Maier, R.V., Mannick, J., McKinley, B., Miller-Graziano, C., Mindrinos, M.N., Minei, J., Moldawer, L.L., Moore, E.,

Moore, F., Munford, R., Nathens, A., O'keefe, G., Purdue, G., Rahme, L., Remick, D., Sailors, M., Schoenfeld, D.A., Shapiro, M., Silver, G., Smith, R., Stephanopoulos, G., Stormo, G., Tompkins, R.G., Toner, M., Warren, S., West, M., Wolfe, S., Xiao, W., Young, V. and Host Response to Injury Large-Scale Collaborative Research Program, 2005. Application of genome-wide expression analysis to human health and disease. Proceedings of the National Academy of Sciences USA 102(13): 4801-4806.

Consortium WTCC, 2007. Genome-wide association study of 14,000 cases of seven common diseases and 3,000 shared controls. Nature 447(7145): 661-678.

Coulier, L., Wopereis, S., Hendriks, H., Radonjic, M. and Jellema, R., in press. Applied chemometrics in systems biology and nutritional metabolomics. Comprehensive Chemometrics: in press.

Crujeiras, A.B., Parra, D., Milagro, F.I., Goyenechea, E., Larrarte, E., Margareto, J. and Martínez, J.A., 2008. Differential expression of oxidative stress and inflammation related genes in peripheral blood mononuclear cells in response to a low-calorie diet: a nutrigenomics study. OMICS: A Journal of Integrative Biology 12(4): 251-261.

Curtis, R.K., Oresic, M. and Vidal-Puig, A., 2005. Pathways to the analysis of microarray data. Trends in Biotechnology 23(8): 429-435.

De Roos, B. and McArdle, H.J., 2008. Proteomics as a tool for the modelling of biological processes and biomarker development in nutrition research. British Journal of Nutrition 99: S66-S71.

De Roos, B., Duivenvoorden, I., Rucklidge, G., Reid, M., Ross, K., Lamers, R.J., Voshol, P.J., Havekes, L.M. and Teusink, B., 2005. Response of apolipoprotein E*3-Leiden transgenic mice to dietary fatty acids: combining liver proteomics with physiological data. FASEB J 19: 813-815.

De Roos, B., Duthie, S.J., Polley, A.C., Mulholland, F., Bouwman, F.G., Heim, C., Rucklidge, G.J., Johnson, I.T., Mariman, E.C., Daniel, H. and Elliott, R.M., 2008a. Proteomic methodological recommendations for studies involving human plasma, platelets, and peripheral blood mononuclear cells. Journal of Proteome Research 7: 2280-2290.

De Roos, B., Geelen, A., Ross, K., Rucklidge, G., Reid, M., Duncan, G., Caslake, M., Horgan, G. and Brouwer, I.A., 2008b. Identification of potential serum biomarkers of inflammation and lipid modulation that are altered by fish oil supplementation in healthy volunteers. Proteomics 8: 1965-1974.

Dragsted, L.O., Krath, B., Ravn-Haren, G., Vogel, U.B., Vinggaard, A.M., Bo Jensen, P., Loft, S., Rasmussen, S.E., Sandstrom, T.B. and Pedersen, A., 2006. Biological effects of fruit and vegetables. Proceedings of the Nutrition Society 65(1): 61-67.

Draisma, H.H., Reijmers, T.H., Bobeldijk-Pastorova, I., Meulman, J.J., Estourgie-Van Burk, G.F., Bartels, M., Ramaker, R., Van der Greef, J., Boomsma, D.I. and Hankemeier, T., 2008. Similarities and differences in lipidomics profiles among healthy monozygotic twin pairs. OMICS: A Journal of Integrative Biology 12(1): 17-31.

Dumas, M.E., Maibaum, E.C., Teague, C., Ueshima, H., Zhou, B., Lindon, J.C., Nicholson, J.K., Stamler, J., Elliott, P., Chan, Q. and Holmes, E., 2006. Assessment of analytical reproducibility of 1H NMR spectroscopy based metabonomics for large-scale epidemiological research: the INTERMAP Study: Analytical Chemistry 78(7): 2199-2208.

Eady, J.J., Wortley, G.M., Wormstone, Y.M., Hughes, J.C., Astley, S.B., Foxall, R.J., Doleman, J.F. and Elliott, R.M., 2005. Variation in gene expression profiles of peripheral blood mononuclear cells from healthy volunteers. Physiological Genomics 22(3):402-411.

Eckhardt, F., Lewin, J., Cortese, R., Rakyan, V.K., Attwood, J., Burger, M., Burton, J., Cox, T.V., Davies, R., Down, T.A., Haefliger, C., Horton, R., Howe, K., Jackson, D.K., Kunde, J., Koenig, C., Liddle, J., Niblett, D., Otto, T., Pettett, R., Seemann, S., Thompson, C., West, T., Rogers, J., Olek, A., Berlin, K. and Beck, S., 2006. DNA methylation profiling of human chromosomes 6, 20 and 22. Nature Genetics 38: 1378-1385.

Fang, X and Zhang, W.W., 2008. Affinity separation and enrichment methods in proteomic analysis. Journal of Proteomics 71: 284-303.

Fraga, M.F., Ballestar, E., Paz, M.F., Ropero, S., Setien, F., Ballestar, M.L., Heine-Suner, D., Cigudosa, J.C., Urioste, M., Benitez, J., Boix-Chornet, M., Sanchez-Aguilera, A., Ling, C., Carlsson, E., Poulsen, P., Vaag, A., Stephan, Z., Spector, T.D., Wu, Y.Z., Plass, C. and Esteller, M., 2005. Epigenetic differences arise during the lifetime of monozygotic twins. Proceedings of the National Academy of Science USA 102: 10604-10609.

Friedman, D.B., Hill, S., Keller, J.W., Merchant, N.B., Levy, S.E., Coffey, R.J. and Caprioli, R.M., 2004. Proteome analysis of human colon cancer by two-dimensional difference gel electrophoresis and mass spectrometry. Proteomics 4(3): 793-811.

Friso, S., Choi, S.-W., Girelli, D., Mason, J.B., Dolnikowski, G.G., Bagley, P.J., Olivieri, O., Jacques, P.F., Rosenberg, I.H., Corrocher, R. and Selhub, J., 2002. A common mutation in the 5,10-methylenetetrahydrofolate reductase gene affects genomic DNA methylation through an interaction with folate status. Proceedings of the National Academy of Science USA 99: 5606-5611.

Fuchs, D., Piller, R., Linseisen, J., Daniel, H. and Wenzel, U., 2007a. The human peripheral blood mononuclear cell proteome responds to a dietary flaxseed-intervention and proteins identified suggest a protective effect in atherosclerosis. Proteomics 7: 3278-3288.

Fuchs, D., Vafeiadou, K., Hall, W.L., Daniel, H., Williams, C.M., Schroot, J.H. and Wenzel, U., 2007b. Proteomic biomarkers of peripheral blood mononuclear cells obtained from postmenopausal women undergoing an intervention with soy isoflavones. American Journal of Clinical Nutriton 86: 1369-1375.

Garcia, A., Watson, S.P., Dwek, R.A. and Zitzmann, N., 2005. Applying proteomics technology to platelet research. Mass Spectrometry Reviews 24:918-930.

Gerszten, R.E. and Wang, T.J., 2008. The search for new cardiovascular biomarkers. Nature 451: 949-52.

Gika, H.G., Theodoridis, G.A. and Wilson, I.D., 2008. Liquid chromatography and ultra-performance liquid chromatography-mass spectrometry fingerprinting of human urine. Sample stability under different handling and storage conditions for metabonomics studies. Journal of Chromatography A 1189: 314-322.

Hake, S.B., Xiao, A. and Allis, C.D., 2004. Linking the epigenetic 'language' of covalent histone modifications to cancer. British Journal of Cancer 90: 761-769.

Horgan, G.W., 2007. Sample size and replication in 2D gel electrophoresis studies. Journal of Proteome Research 6(7): 2884-2887.

Hudgin, R.L., Pricer, W.E. Jr, Ashwell, G., Stockert, R.J. and Morell, A.G., 1974. Isolation and properties of a rabbit liver binding-protein specific for asialoglycoproteins. Journal of Biological Chemistry 249(17): 5536-5543.

Hunt, S.M., Thomas, M.R., Sebastian, L.T., Pedersen, S.K., Harcourt, R.L., Sloane, A.J. and Wilkins, M.R., 2005. Optimal replication and the importance of experimental design for gel-based quantitative proteomics. Journal of Proteome Research 4: 809-819.

Ilveskoski, E., Perola, M., Lehtimäki, T., Laippala, P., Savolainen, V., Pajarinen, J., Penttilä, A., Lalu, K.H., Männikkö, A., Liesto, K.K., Koivula, T. and Karhunen, P.J., 1999. Age-dependent association of apolipoprotein E genotype with coronary and aortic atherosclerosis in middle-aged men: an autopsy study. Circulation 100(6):608-613.

Irizarry, R.A., Warren, D., Spencer, F., Kim, I.F., Biswal, S., Frank, B.C., Gabrielson, E., Garcia, J.G., Geoghegan, J., Germino, G., Griffin, C., Hilmer, S.C., Hoffman, E., Jedlicka, A.E., Kawasaki, E., Martínez-Murillo, F., Morsberger, L., Lee, H., Petersen, D., Quackenbush, J., Scott, A., Wilson, M., Yang, Y., Ye, S.Q. and Yu, W., 2005. Multiple-laboratory comparison of microarray platforms. Nature Methods 2(5): 345-350.

Issaq, H.J., Abbott, E. and Veenstra, T.D., 2008. Utility of separation science in metabolomic studies. Journal of Separation Science 31: 1936-1947.

Jacob, R.A., Gretz, D.M., Taylor, P.C., James, S.J., Pogribny, I.P., Miller, B.J., Henning, S. M. and Swendseid, M.E., 1998. Moderate folate depletion increases plasma homocysteine and decreases lymphocyte DNA methylation in postmenopausal women. Journal of Nutrition 128: 1204-1212.

Jirtle, R.L. and Skinner, M.K., 2007. Environmental epigenomics and disease susceptibility. Nature Reviews Genetics 8: 253-262.

Joost, H.G., Gibney, M.J., Cashman, K.D., Görman, U., Hesketh, J.E., Muller, M., Van Ommen, B., Williams, C.M. and Mathers, J.C., 2007. Personalised nutrition: status and perspectives. British Journal of Nutrition 98: 26-31.

Kendziorski, C., Irizarry, R.A., Chen, K.-S., Haag, J.D. and Gould, M.N., 2005. On the utility of pooling biological samples in microarray experiments. Proceedings of the National Academy of Sciences USA 102(12): 4252-4257.

Klerk, M., Verhoef, P., Clarke, R., Blom, H.J., Kok, F.J. and Schouten, E.G., 2002. MTHFR 677C-->T polymorphism and risk of coronary heart disease: a meta-analysis. JAMA 288(16): 2023-2031.

Koek, M.M.; Muilwijk, B.; Van der Werf, M.J. and Hankemeier, T.H., 2006. Microbial metabolomics with gas chromatography/mass spectrometry. Analytical Chemistry 78: 1272-1281.

Korshunova, Y., Maloney, R.K., Lakey, N., Citek, R.W., Bacher, B., Budiman, A., Ordway, J.M., McCombie, W.R., Leon, J., Jeddeloh, J.A. and McPherson, J.D., 2008. Massively parallel bisulphite pyrosequencing reveals the molecular complexity of breast cancer-associated cytosine-methylation patterns obtained from tissue and serum DNA. Genome Research 18: 19-29.

Lasky-Su, J., Lyon, H.N., Emilsson, V., Heid, I.M., Molony, C., Raby, B.A., Lazarus, R., Klanderman, B., Soto-Quiros, M.E., Avila, L., Silverman, E.K., Thorleifsson, G., Thorsteinsdottir, U., Kronenberg, F., Vollmert, C., Illig, T., Fox, C.S., Levy, D., Laird, N., Ding, X., McQueen, M.B., Butler, J., Ardlie, K., Papoutsakis, C., Dedoussis, G., O'Donnell, C.J., Wichmann, H.E., Celedón, J.C., Schadt, E., Hirschhorn, J., Weiss, S.T., Stefansson, K. and Lange, C., 2008. On the replication of genetic associations: timing can be everything! American Journal of Human Genetics 82(4): 849-858.

Macaulay, I.C., Carr, P., Gusnanto, A., Ouwehand, W.H., Fitzgerald, D. and Watkins, N.A., 2005. Platelet genomics and proteomics in human health and disease. Journal of Clinical Investigation 115: 3370-3377.

Martin, K.J., Graner, E., Li Y., Price, L.M., Kritzman, B.M., Fournier, M.V., Rhei, E. and Pardee, A.B., 2001. High-sensitivity array analysis of gene expression for the early detection of disseminated breast tumor cells in peripheral blood. Proceedings of the National Academy of Sciences USA 98(5): 2646-2651.

Mathers, J.C., 2006a. Candidate mechanisms for interactions between nutrients and genes. In: Choi, S.-W. and Friso, S. (eds.), Nutrient-gene interactions in cancer. Taylor & Francis Group, Boca Raton, 19-36pp.

Mathers, J.C., 2006b. Nutritional modulation of ageing: genomic and epigenetic approaches. Mechanisms of Ageing and Development 127: 584-589.

Mckay, J., Adriaens, M., Ford, D., Relton, C., Evelo, C. and Mathers, J.C., 2009. Bioinformatic interrogation of expression array data to identify nutritionally regulated genes potentially modulated by DNA methylation – in press

Minihane, A.M., Khan, S., Leigh-Firbank, E.C., Talmud, P., Wright, J.W., Murphy, M.C., Griffin, B.A. and Williams, C.M., 2000. ApoE polymorphism and fish oil supplementation in subjects with an atherogenic lipoprotein phenotype. Arteriosclerosis, Thrombosis, and Vascular Biology 20(8):1990-1997.

Müller, M. and Kersten, S., 2003. Nutrigenomics: Goals and strategies. Nature Reviews Genetics 4: 315-322.

National Institutes of Health, 2008. NIH announces new initiative in epigenomics. NIH News: 22 January 2008.

Nijhout, H.F., Reed, M.C., Anderson, D.F., Mattingly, J.C., James, S.J. and Ulrich, C.M., 2006. Long-range allosteric interactions between the folate and methionine cycles stabilize DNA methylation reaction rate. Epigenetics 1: 81-87.

O'Neill, E.E., Brock, C.J., Von Kriegsheim, A.F., Pearce, A.C., Dwek, R.A., Watson, S.P. and Hebestreit, H.F., 2002. Towards complete analysis of the platelet proteome. Proteomics 2: 288-305.

Orchard, S., Hermjakob, H. and Apweiler, R., 2003. The proteomics standards initiative. Proteomics 3(7): 1374-1376.

Ordovas, J.M., 2002. Gene-diet interaction and plasma lipid responses to dietary intervention. Biochemical Society Transactions 30(2): 68-73.

Ordovas, J.M., Corella, D., Demissie, S., Cupples, L.A., Couture, P., Coltell, O., Wilson, P.W., Schaefer, E.J. and Tucker, K.L., 2002. Dietary fat intake determines the effect of a common polymorphism in the hepatic lipase gene promoter on high-density lipoprotein metabolism: evidence of a strong dose effect in this gene-nutrient interaction in the Framingham Study. Circulation 106(18): 2315-2321.

Pearson, T.A. and Manolio, T.A., 2008. How to interpret a genome-wide association study. JAMA 299(11): 1335-1344.

Polley, A.C., Mulholland, F., Pin, C., Williams, E.A., Bradburn, D.M., Mills, S.J., Mathers, J.C. and Johnson I.T., 2006. Proteomic analysis reveals field-wide changes in protein expression in the morphologically normal mucosa of patients with colorectal neoplasia. Cancer Research 66(13): 6553-6562.

Ptitsyn, A.A., Zvonic, S., Conrad, S.A., Scott, L.K., Mynatt, R.L. and Gimble, J.M., 2006. Circadian clocks are resounding in peripheral tissues. PLoS Computational Biology 2(3): e16. doi:10.1371/journal.pcbi.0020016

Radich, J.P., Mao, M., Stepaniants, S., Biery, M., Castle, J., Ward, T., Schimmack, G., Kobayashi, S., Carleton, M., Lampe, J. and Linsley, P.S., 2004. Individual-specific variation of gene expression in peripheral blood leukocytes. Genomics 83(6): 980-988.

Rakyan, V.K., Hildmann, T., Novik, K.L., Lewin, J., Tost, J., Cox, A.V., Andrews, T.D., Howe, K.L., Otto, T., Olek, A., Fischer, J., Gut, I. G., Berlin, K. and Beck, S., 2004. DNA methylation profiling of the human major histocompatibility complex: a pilot study for the human epigenome project. PLoS Biology 2(12): e405. doi:10.1371/journal.pbio.0020405

Reik, W., Dean, W. and Walter, J., 2001. Epigenetic reprogramming in mammalian development. Science 293: 1089-1093.

Sangster, T., Major, H., Plumb, R., Wilson, A.J. and Wilson, I.D., 2006. A pragmatic and readily implemented quality control strategy for HPLC-MS and GC-MS-based metabolomic analysis. Analyst 131: 1075-1078.

Saude, E.J. and Sykes, B.D., 2007. Urine stability for metabolomic studies: effects of preparation and storage. Metabolomics 3: 19-27.

Schnackenberg, L.K., Kaput, J. and Beger, R.D., 2008. Metabolomics: a tool for personalizing medicine? Biotechnology Advances 26(2): 169-176.

Shaham, O., Wei, R., Wang, T.J., Ricciardi, C., Lewis, G.D., Vasan, R.S., Carr, S.A., Thadhani, R., Gerszten, R.E. and Mootha, V.K., 2008. Metabolic profiling of the human response to a glucose challenge reveals distinct axes of insulin sensitivity. Molecular Systems Biology 4: 214.

Slupsky, C.M., Rankin, K.N., Wagner, J., Fu, H., Chang, D., Weljie, A.M., Saude, E.J., Lix, B., Adamko, D.J., Shah, S., Greiner, R., Sykes, B.D. and Marrie, T.J., 2007. Investigations of the effects of gender, diurnal variation, and age in human urinary metabolomic profiles. Analytical Chemistry 79(18): 6995-7004.

Smilde, A.K., Bro, R. and Geladi, P., 2004. Multi-way analysis: Applications in the chemical sciences. Wiley, West Sussex, UK, pp. 221-256.

Song, Y., Stampfer, M.J. and Liu, S., 2004. Meta-analysis: apolipoprotein E genotypes and risk for coronary heart disease. Annals of Internal Medicine 141(2): 137-147.

Stulík, J., Koupilova, K., Osterreicher, J., Knízek, J., Macela, A., Bures, J., Jandík, P., Langr, F., Dedic, K. and Jungblut, P.R., 1999. Protein abundance alterations in matched sets of macroscopically normal colon mucosa and colorectal carcinoma. Electrophoresis 20(18): 3638-3646.

Taylor, C.F., Paton, N.W., Lilley, K.S., Binz, P.A., Julian, R.K. Jr, Jones, A.R., Zhu, W., Apweiler, R., Aebersold, R., Deutsch, E.W., Dunn, M.J., Heck, A.J., Leitner, A., Macht, M., Mann, M., Martens, L., Neubert, T.A., Patterson, S.D., Ping, P., Seymour, S.L., Souda, P., Tsugita, A., Vandekerckhove, J., Vondriska, T.M., Whitelegge, J.P., Wilkins, M.R., Xenarios, I., Yates, J.R. 3rd and Hermjakob, H., 2007. The minimum information about a proteomics experiment (MIAPE). Nature Biotechnology 25(8): 887-893.

Thissen, U., Wopereis, S., Van den Berg, S.A., Bobeldijk, I., Kleemann, R., Kooistra, T., van Dijk, K.W., Van Ommen, B. and Smilde, A.K., 2009. Improving the analysis of designed studies by combining statistical modelling with study design information. BMC Bioinformatics 10: 52.

Valk, P.J., Verhaak, R.G., Beijen, M.A., Erpelinck, C.A., Barjesteh van Waalwijk van Doorn-Khosrovani, S., Boer, J.M., Beverloo, H.B., Moorhouse, M.J., Van der Spek, P.J., Löwenberg, B. and Delwel, R., 2004. Prognostically useful gene-expression profiles in acute myeloid leukemia. New England Journal of Medicine 350(16): 1617-1628.

Van der Greef, J., Hankemeier, T. and McBurney, R.N., 2006. Metabolomics-based systems biology and personalized medicine: moving towards n = 1 clinical trials? Pharmacogenomics 7(7): 1087-1094.

Van der Greef, J., Martin, S., Juhasz, P., Adourian, A., Plasterer, T., Verheij, E.R. and McBurney, R.N., 2007. The art and practice of systems biology in medicine: Mapping patterns of relationships. Journal of Proteome Research 6: 1540-1559.

Van der Greef, J., Martin, S., Juhasz, P., Adourian, A., Plasterer, T., Verheij, E.R. and McBurney, R.N., 2007. The art and practice of systems biology in medicine: mapping patterns of relationships. Journal of Proteome Research 6: 1540-1559.

Van Erk, M.J., Blom, W.A., Van Ommen, B. and Hendriks, H.F., 2006. High-protein and high-carbohydrate breakfasts differentially change the transcriptome of human blood cells. American Journal of Clinical Nutrition 84: 1233-1241.

Walsh, M.C., Brennan, L., Malthouse, J.P., Roche, H.M. and Gibney, M.J., 2006. Effect of acute dietary standardization on the urinary, plasma, and salivary metabolomic profiles of healthy humans: American Journal of Clinical Nutrition 84(3): 531-539.

Wheelock, A.M. and Goto, S., 2006. Effects of post-electrophoretic analysis on variance in gel-based proteomics. Expert Reviews in Proteomics 3: 129-142.

Whitney, A.R., Diehn, M., Popper, S.J., Alizadeh, A.A., Boldrick, J.C., Relman, D.A. and Brown, P.O., 2003. Individuality and variation in gene expression patterns in human blood. Proceedings of the National Academy of Sciences USA 100(4): 1896-1901.

Wilkins, M.R., Appel, R.D., van Eyk, J.E., Chung, M.C., Görg, A., Hecker, M., Huber, L.A., Langen, H., Link, A.J., Paik, Y.K., Patterson, S.D., Pennington, S.R., Rabilloud, T., Simpson, R.J., Weiss, W. and Dunn, M.J., 2006. Guidelines for the next 10 years of proteomics. Proteomics 6: 4-8.

Wilson, A.S., Power, B.E. and Molloy, P.L., 2007 DNA hypomethylation and human diseases. Biochimica et Biophysica Acta 1775: 138-1362.

Wopereis, S., Rubingh, C.M., Van Erk, M.J., Verheij, E.R., van Vliet, T., Cnubben, N.H., Smilde, A.K., Van der Greef, J., Van Ommen, B. and Hendriks, H.F., 2009. Metabolic profiling of the response to an oral glucose tolerance test detects subtle metabolic changes. PLoS ONE 4(2):e4525 doi:10.1371/journal.pone.0004525.

Yang, A.S., Estecio, M.R., Doshi, K., Kondo, Y., Tajara, E.H. and Issa, J.P., 2004. A simple method for estimating global DNA methylation using bisulfite PCR of repetitive DNA elements. Nucleic Acids Research 32: e38.

Zdravkovic, S., Wienke, A., Pedersen, N.L., Marenberg, M.E., Yashin, A.I. and De Faire, U., 2002. Heritability of death from coronary heart disease: a 36-year follow-up of 20, 966 Swedish twins. Journal of Internal Medicine 252(3): 247-254.

Zvonic, S., Ptitsyn, A.A., Conrad, S.A., Scott, L.K., Floyd, Z.E., Kilroy, G., Wu, X., Goh, B.C., Mynatt, R.L. and Gimble, J.M., 2006. Characterization of peripheral circadian clocks in adipose tissues. Diabetes 55(4): 962-970.

Index